# THERMODYNAMICS
## Ideal Gas Basic Theory

U. KIVI

# THERMODYNAMICS
## Ideal Gas Basic Theory

FSC
www.fsc.org
MIX
Paperi vastuul -
lisista lähteistä
Paper from
responsible sources
FSC® C105338



Book design and Cover: Books on Demand

Publisher: BoD – Books on Demand, Helsinki, Finland

Print: BoD – Books on Demand, Norderstedt, Germany

ISBN: 978-952-80-7741-1

# CONTENT

# FOREWORD

Coffee, roll and sudoku in swimming hall cafe or in some other common cafe constitute an essential part of my swimming hoppy. Sudoku was gradually replaced by Einstein's relativity theory calculations about five years ago. One reason for the change was that a family with children moved to the residence beneath my residence. The children made unbearable noise every evening, so I began to hang in cafes late almost every evening and calculated physics. In couple of years the noise was reduced essentially – due to complaints of some of my neighbors. But I continued with my new calculation hoppy, now increasingly also at home. I had of course been interested in relativity theory and other fields of physics also before, but my main action was only to read appropriate literature. To my surprise I could attain new results, and published the results in a book (U. Kivi 'Painovoima, Newtonin ja Einsteinin teoria helpolla tavalla'and the English translation U. Kivi: 'Gravitation, Exact calculation of Newton and Einstein theory'). I don't know yet whether the results presented in the 'Gravitation' book are worth anything, but if they are, they are not my first achievements in the field of physics/mathematics. Although I have published nothing, I obtained some remarkable accomplishments already in 80's, when I began my career in nuclear field. I have listed my most important achievements in appendix 1 as labours of Joracles. This book is linked with the labour number 1. As an initiator and adviser for that labour acted my boss of that time. As a theoretical physicist and mathematician I could formulate his ideas got by examining steam tables into form of mathematical algorithms. As a result we got a new way to calculate the pressure of air-steam-water –mixture. The method lies on a very firm ground, because it is based on the laws of conservation of mass and energy. Success creates success and encouraged by this achievement and supported by my boss I started to solve also other mathematical problems in my job domain and achieved good result as is presented in appendix 1.

Without those successes presented in appendix 1 I would hardly have had the courage to begin to solve the equations of general relativity. Relativity theory solutions do not have such economic significance as the mathematical algorithms presented in appendix 1, but as an a bit original view of solving gravitation the theory presented in the 'Gravitation' -book hopefully promotes the development of physics as a science (or maybe not if it turns out to be pure grab). Inspired by the 'Gravitation' –book I decided to write also a book about thermodynamics. Relativity theory does not have any connection to my profession, but with thermodynamics I spent a large part of my first professional years.

As a student I was a bit disappointed to notice that my understanding of the fundamental structure of classical thermodynamics was rather weak after the course of statistical physics at Technical High School. As I began my professional career, one of my first tasks was to examine the pressure and temperature behavior of Loviisa nuclear power station containment in an accident that is called loss of coolant accident. In this accident the main circulation pipe of reactor coolant water is broken and hot water and steam discharge into the containment. This is typical problem in the field of thermodynamics. Of course this situation had been analyzed with the help of computer calculations in the design materials of the power station, but at that time even individual computer runs were expensive, so we decided with my boss to make our own computer program for thermodynamic calculations. The program was completed and we could vary among other things the size of the leak, the amount of heat flow into structures, function of ice condenser and the function of the containment spray system. With the help of the program we could examine the accident situation more thoroughly than was done in design material and we could also verify the design calculation results. Our calculations indicated that the design calculations had been performed in a very firm way or conservatively, if we use our professional term.

Because the theory of thermodynamics was not easy to understand for me (I suppose that it was not easy for almost anybody), I had to deliberate also the foundations of the theory. Often the most efficient way to understand the foundations of some theory is to consider the simplest possible model, which in the case of thermodynamics is the ideal gas model. I present in this book the ideas that I got when making the thermodynamics program and also some ideas that matured slowly at later years. The ideas are mainly applied for the ideal gas calculations.

Helsinki 31.1.2020
Joracles alias Uuno Kivi alias Jorma Rantakivi

# 1. INTRODUCTION

Thermodynamics deals with properties and relations of quantities known to everybody (temperature, pressure, density). The need to understand the function of thermal power machines (such as steam engine) gave the impulse to develop the theory. Energy was found to be the fundamental quantity of the theory and it governs the thermodynamic properties of mater in form of heat or mechanical energy. Thermal power machine is able to convert heat to mechanical energy. The theory of thermodynamics was needed to optimize the function of the thermal power machines in such a way that the gain of mechanical energy would be maximized. It turned out that the quantities introduced above were not enough to create a functional theory, but additionally a new more abstract quantity was needed to describe the thermodynamical state of a system. The new quantity was named entropy. The theory of thermodynamics and kinetic theory of gases was developed in 1800 century and at the beginning of 1900 century. The new quantity – entropy – and its physical interpretation had a central role in that development. Entropy turned out to be a measure for the disorder of a system. The smaller the entropy the more ordered the system is. Because physical processes have a natural tendency to go towards higher disorder, entropy was found always to increase or stay constant. Entropy may decrease in a subsystem of a larger system, but if one considers the entropy of the whole system, it either increases or stays constant. If we for example have two containers filled with gas and having different temperatures in contact with each other, heat flows from the gas with higher temperature to the gas with lower temperature and the entropy of that gas that is cooling decreases, but this is compensated by entropy increase of the gas that warms up.

I was especially irritated by the fact that you did not know the functional relationships between the different quantities even in the case of ideal gas thermodynamics. Only relationships of certain partial derivatives of the quantities were available. Because of this the theory in a way hanged in the air and mastering the theory seemed to demand a long-term practice in manipulating the appropriate formulas. So I tried solemnly to develop functional formulas between the different quantities to make the theory more concrete. I succeeded in doing this and it was surprisingly easy and the functional relationships are presented in chapter 5 'Ideal gas'. Now I have to say that I know the thermodynamic literature so poorly that I cannot claim that these formulas would be quite new, but they seem to be missing at least in the older thermodynamic textbooks.

In chapter 6 'Ideal gas thermodynamic processes' I construct the needed calculation formulas and describe the development of processes, when two of quantities pressure, volume, temperature or heat transfer are given as functions of time. I hope that this chapter helps especially those people (for example students) that begin to familiarize themselves to thermodynamics. A thermodynamic process is calculated step by step using small time steps. A special feature of the method used in this book is that the two quantities given as functions time are allowed to change simultaneously. Normally the calculation is performed so that one of the quantities is kept constant while changing the other one. The result depends on the value of the time step and to get an accurate result you have to use a small enough time steps. One may assume that the dependence on the value of the time step is not so strong with the method used in this book compared to the normal method. If the dependence of the given quantities on time is not linear, the desired accuracy may be reached by using small enough time step. A Fortran program is listed in appendix 2 with the help of which you may calculate the state of an ideal gas system, when two of quantities pressure, volume, temperature or heat transfer are given as functions of time. You may choose two out of four in six ways so there are six different calculation cases. An example calculation is performed for each case. The result is given as graphs and an explanation for each graph is given. As the last case the program is used to calculate the famous Carnot process.

# 2. THE BASIC PROBLEM

Originally the laws obeyed by the ideal gas were constructed as a result of many experiments and physical theories to explain these laws were constructed later. These theories are out of the scope of this book so that we take the ideal gas laws as given. Ideal gas has been found to obey the following equations

(2.1) $$pV=nkT$$

(2.2) $$U=\alpha nkT$$

where p is pressure, V is volume, n is the number of molecules, k is Boltzmann constant, U is the inner energy, α is molecular heat capacity = 1.5 for one atomic gases and = 2.5 for two atomic gases and T is the temperature.

If we leave the mass transfer out of consideration a thermodynamic process may take place in two different ways: you may either add energy to the system by heat transfer or you may change the volume of the system by compression. In the later case in addition to the volume change the inner energy of the system increases by the compression energy ($-p\,\Delta V$). The basic problem of thermodynamics is that, if the state of the system is known at time 0 ($p_0$,$V_0$,$T_0$) and if an amount $\Delta Q$ of heat energy is added to the system and the system volume is compressed by an amount $\Delta V$, the state of the system is not uniquely defined by this information, but is depended on the order of volume change and heat energy addition.

The end state is a unique function of the volume and the inner energy at the end of the process. The issue is that, if the heat energy is added in the initial stage of the compression process, one compresses hotter gas and gas with higher pressure, so the compression work which is added to the inner energy is larger. The result is that the inner energy and the temperature are higher compared to a case in which the heat energy is added at the end stage of the compression process. The coolest gas and the gas with smallest pressure is achieved, if all heat energy is added only after the compression process.

To solve this problem a new quantity – entropy – was introduced in 1800 century. Entropy (S) was defined in such a way that the end state of a compression process of a system is uniquely defined, if you know the volume change ΔV and the entropy change ΔS of the system, independent of the time moment of heat energy addition to the system.

# 3. ENTROPY

The entropy change of a system is defined as follows

(3.1) $$dS = dQ/T$$

where dS is the entropy change, dQ is the heat energy addition to the system and T is the temperature of the system. You may also write

(3.2) $$dS = dU/T$$

if only heat energy addition is changing the inner energy U (no volume changes). Inserting formula 2.2 into formula 3.2 we get

(3.3) $$dS = \alpha nk\, dT/T$$

Integrate $S_0$ -> S and $T_0$ -> T, you get

(3.4) $$S - S_0 = \alpha nk \ln(T/T_0)$$

It is tempting to choose value zero for entropy at the absolute temperature zero point (that is $S_0 = 0$, $T_0 = 0$). The right side of formula 3.4 has value infinity, when temperature $T_0 = 0$, but when T approaches $T_0$, S approaches $S_0$ , so that choice $S_0 = 0$, $T_0 = 0$ is probably possible at least as a limit value. With the help of formula 2.2 you may write formula 3.4 as

(3.5) $$S - S_0 = \alpha nk \ln(U/U_0)$$

which gives the relationship between the inner energy and entropy. You may change the inner energy of the gas with an arbitrary value by changing the volume of the gas and entropy stays constant, because according to formula 3.1 the value of entropy depends only on the inserted heat energy. So the inner energy may have whatever value with certain entropy, but formula 3.5 may be considered as the natural relationship between entropy and inner energy. If the value of the entropy differs from the value given by formula 3.5 the gas has been either compressed or it has expanded.

So entropy doesn't depend naturally on volume. However in chapter 5 we derive a formula for entropy as a function of inner energy and volume. This is to be understood so that, because in a thermodynamic process many quantities change simultaneously, it may appear that entropy depends on volume, but this in reality not the case. If for example you compress a gas so that the inner energy is constant, you have to export the compression energy as heat transfer out of the gas. It seems that entropy is changing as a function of volume, but actually the entropy change is caused by the heat transfer.

Next we derive the dependence between volume and inner energy, when there is no heat transfer (so that entropy stays constant). The change of inner energy caused by the compression may be written as

$$dU = -pdV = -(U/(\alpha V))dV \tag{3.6}$$

So that

$$(\alpha/U)dU = -dV/V \tag{3.7}$$

Integrate $U_0$ -> U and $V_0$ -> V and you get

$$\frac{V_0}{V} = \left(\frac{U}{U_0}\right)^\alpha = \left(\frac{T}{T_0}\right)^\alpha \tag{3.8}$$

By inserting the ideal gas formulas

$$V = \frac{nkT}{p} \tag{3.9}$$

and

$$V_0 = \frac{nkT_0}{p_0} \tag{3.10}$$

you get the dependency of pressure on temperature, when gas is compressed without heat transfer

(3.11) $$\frac{p}{p_0} = \left(\frac{U}{U_0}\right)^{\alpha+1} = \left(\frac{T}{T_0}\right)^{\alpha+1}$$

# 4. BASICS

In principle the inner energy of a system may increase in two different ways. You may add heat energy to the system through heat transfer - in this case the inner energy of the system increases by the amount of the added heat – or you may compress the system. In this case the inner energy increases by the amount of compression work (force multiplied by distance or pressure multiplied by the volume change). This may be expressed by the formula

(4.1) $$dU = dQ-pdV$$

where dU is the change in inner energy, dQ is the added heat energy by heat transfer and -pdV is the compression work done to the system.

Next we write the heat transfer in the following way

(4.2) $$dQ = TdS$$

where dS is the entropy added to the system. This is only a change of variable so the entropy added to the system corresponds to the energy added to the system. If added heat is zero that is dQ = 0, the entropy of the system stays constant that is dS = 0, as you may calculate from formula 4.2.

Thermodynamics is constructed for so called reversible processes that obey formulas 4.1 and 4.2. It is true that the compression may be done in a harsh way so that system gains more energy than -pdV. Consider an ideal gas. At room temperature the speed of the gas molecules varies between 100 - 1000 m/s depending on the weight of the molecules. If you compress gas in cylinder with a piston the velocity of which is lower than the average gas molecule velocity, the molecules scatter elastically with the piston gaining energy (a molecule with speed v scatters back from the piston with speed $v+2v_m$, where $v_m$ is the speed of the piston) and the inner energy of the system changes with amount -pdV. However you may add an arbitrary large amount of energy to the gas by making the compression very quickly. In this case the inner energy of the gas increases more than

–pdV and formula 4.1 is not valid. This kind of a process is called irreversible. Because thermodynamics has been constructed so that that the change of inner energy caused by volume change is always –pdV one reaches a dead end. This 'dead end' is solved so that one separates the normal compression work -pdV and the additional work done by the harsh compression. One does not anymore demand that the energy used in formula 4.2 should be only due to heat transfer, but if one brings in a compression more energy than –pdV, it can be used in formula 4.2 in the same way as heat energy. So irreversible compression can be divided into reversible compression plus extra energy addition. The energy addition may be treated in the same way as added heat energy. In this way the entropy changes in an irreversible process, although there is no heat addition, but the reason for entropy change is quite the same as in the case of heat transfer, more energy than amount -pdV is transferred into the system. We notice that energy is added to the system in an irreversible compression so that entropy increases (formula 4.2).

Next we consider expansion of an ideal gas. If piston moves with a speed that is lower than the speed of gas molecules that is under the speed 100-1000 m/s in room temperature, gas molecules have time to follow the piston and collide with the piston losing energy (a molecule with speed v scatters back from the piston with speed $v-2*v_m$, where $v_m$ is the speed of the piston) and the inner energy of the system decreases with amount -pdV. But if the piston moves faster than the average speed of gas molecules, less and less gas molecules are able to collide with the piston to lose energy and the inner energy of the system decreases less than the amount -pdV. As an extreme case the piston is moved to the new position so fast that the inner energy does not change at all (gas expansion to vacuum). Because thermodynamics is constructed so that inner energy is always thought to change in a volume change according to formula 4.1 we arrive a dead end. Again we have to divide the irreversible process (now expansion) into a reversible expansion plus energy transfer, that changes entropy according to formula 4.2. It is in a way funny that also in this opposite action of compression one has to bring energy into the system and entropy increases. This is so because of the way the thermodynamics is constructed. Inner energy is always thought to change according to formula 4.1 (-pdV) as volume changes. So if an 'ugly' volume change does not cause this energy change, you have to import energy to the system as a compensation (although energy is not really added but the energy change caused by the volume change differs from the anticipated value -pdV) and entropy changes according to formula 4.2. In an irreversible expansion inner energy changes less than amount -pdV, so the irreversible expansion may be divided into a reversible expansion plus an energy addition that increases entropy according to formula 4.2.

Both in irreversible compression and in irreversible expansion entropy increases (so entropy increases always!). We may state that the interpretation of both the absolute value of entropy and the change of entropy are matters of interpretation. You may ask what part of the inner energy is interpreted as the original amount of inner energy, what part has been imported by heat transfer (causing an entropy change) and what part of the inner energy has been created by compression (does not cause entropy change).

# 5. IDEAL GAS

It has been found out experimentally that ideal gas obeys the following equations

(5.1) $$pV = nkT$$

(5.2) $$U = \alpha nkT$$

where p is pressure, V is volume, n is the number of molecules, k is Boltzmann constant, U is inner energy, α = molecular heat capacity = 1.5 for one atomic gases and = 2.5 for two atomic gases and T is temperature.

Next we write the change of inner energy with the help of formulas 4.1 and 4.2 into form

(5.3) $$dU = TdS - pdV$$

If U is presented as a function of S and V, its differential may be presented as

(5.4) $$dU=U_S\, dS+U_V\, dV$$

By comparing formulas 5.3 and 5.4 we see that when U is given as a function of S and V (U = U(S,V)) the following partial differential equations are valid

(5.5) $$U_S = \frac{\partial U}{\partial S} = T$$

(5.6) $$U_V = \frac{\partial U}{\partial V} = -p$$

With the help of formula 5.2 formula 5.5 may be written as

(5.7) $$\frac{\partial U}{U} = \frac{\partial S}{\alpha nk},$$

Integrating both sides ($U(S_0,V)$ -> $U(S,V)$ and $S_0$ -> S) we get

(5.8) $$U(S,V) = U(S_0,V)e^{\frac{S-S_0}{\alpha nk}}$$

Using formulas 5.1, 5.2 and 5.6 we may write

(5.9) $$\frac{\partial U}{\partial V} = -\frac{U}{\alpha V}$$

Next we insert U from formula 5.8 into 5.9 and get

(5.10) $$\frac{\partial U(S_0,V)}{U(S_0,V)} = -\frac{\partial V}{(\alpha V)}$$

and integrating both sides ($U(S_0,V_0)$ -> $U(S_0,V)$ and $V_0$ ->V) we get

(5.11) $$U(S_0,V) = U(S_0,V_0)\left(\frac{V}{V_0}\right)^{-\frac{1}{\alpha}}$$

Insert formula 5.11 into 5.8 and you get U as a function of S and V

(5.12) $$U(S,V) = U(S_0,V_0)\left(\frac{V}{V_0}\right)^{-\frac{1}{\alpha}} e^{\frac{S-S_0}{\alpha nk}}$$

Separating the constant term we get

(5.13) $$U(S,V) = \kappa * (V)^{-\frac{1}{\alpha}} e^{\frac{S}{\alpha nk}}$$

where the value of the constant

(5.14) $$\kappa = U(S_0,V_0)\left(\frac{1}{V_0}\right)^{-\frac{1}{\alpha}} e^{-\frac{S_0}{\alpha nk}}$$

depends on the value chosen for the inner energy at point $(S_0,V_0)$ .

By inserting formula 5.13 into formulas 5.5 and 5.6, we get T and p as functions of S and V and using formulas

(5.15) $$H = U + pV,$$

(5.16) $$F = U - ST,$$

(5.17) $$G = U + pV - ST$$

we may write enthalpy (H), Helmholtz free energy (F) and Gibbs free energy (G) as functions of S and V. Next we write all formulas together.

Free variables S and V

(5.18) $$p = \frac{\kappa}{\alpha} e^{\frac{S}{\alpha nk}} V^{-1-1/\alpha}$$

(5.19) $$T = \frac{\kappa}{\alpha nk} e^{\frac{S}{\alpha nk}} V^{-1/\alpha}$$

(5.20) $$U = \kappa\, e^{\frac{S}{\alpha nk}} V^{-1/\alpha}$$

(5.21) $$H = \kappa(\frac{\alpha+1}{\alpha})\, e^{\frac{S}{\alpha nk}} V^{-1/\alpha}$$

(5.22) $$F = \kappa(1 - \frac{S}{\alpha nk})\, e^{\frac{S}{\alpha nk}} V^{-1/\alpha}$$

(5.23) $$G = \kappa(1 + 1/\alpha - \frac{S}{\alpha nk})\, e^{\frac{S}{\alpha nk}} V^{-1/\alpha}$$

Using formulas 5.18 and 5.19 you may calculate that Maxwell equation

(5.24) $$\left(\frac{\partial T}{\partial V}\right)_S = -\left(\frac{\partial p}{\partial S}\right)_V$$

is valid.

You may choose for free variables any pair of variables S, V, p, T. Solve the not free variables from equations 5.18 and 5.19 as functions of the free variables and insert into equations 5.20 – 5.23, then you get U, H, F and G as functions of the chosen free variables. Formulas for all free variable pairs are presented below

Free variables S and T

(5.25) $$p = \frac{\kappa}{\alpha}\, e^{\frac{S}{\alpha nk}} \left(\frac{\kappa e^{\frac{S}{\alpha nk}}}{\alpha nkT}\right)^{-1-\alpha}$$

(5.26) $$V = \left(\frac{\kappa e^{\frac{S}{\alpha nk}}}{\alpha nkT}\right)^{\alpha}$$

(5.27) $$U = \alpha nkT$$

(5.28) $$H = (\alpha + 1)nkT$$

(5.29) $$F = \alpha nkT - ST$$

(5.30) $$G = (\alpha + 1)nkT - ST$$

Free variables S and p

(5.31) $$V = \left\{\frac{p\alpha}{\kappa e^{\frac{S}{\alpha nk}}}\right\}^{\frac{-\alpha}{\alpha+1}}$$

(5.32) $$T = \frac{\kappa e^{\frac{S}{\alpha nk}}}{\alpha nk} \left\{\frac{p\alpha}{\kappa e^{\frac{S}{\alpha nk}}}\right\}^{\frac{1}{\alpha+1}}$$

(5.33) $$U = \kappa e^{\frac{S}{\alpha nk}} \left\{\frac{p\alpha}{\kappa e^{\frac{S}{\alpha nk}}}\right\}^{\frac{1}{\alpha+1}}$$

(5.34) $$H = \frac{\alpha+1}{\alpha} \kappa e^{\frac{S}{\alpha nk}} \left\{\frac{p\alpha}{\kappa e^{\frac{S}{\alpha nk}}}\right\}^{\frac{1}{\alpha+1}}$$

(5.35) $$F = \left(1 - \frac{S}{\alpha nk}\right) \kappa e^{\frac{S}{\alpha nk}} \left\{\frac{p\alpha}{\kappa e^{\frac{S}{\alpha nk}}}\right\}^{\frac{1}{\alpha+1}}$$

(5.36) $$G = \left(1 + \frac{1}{\alpha} - \frac{S}{\alpha nk}\right) \kappa e^{\frac{S}{\alpha nk}} \left\{\frac{p\alpha}{\kappa e^{\frac{S}{\alpha nk}}}\right\}^{\frac{1}{\alpha+1}}$$

With the help of formulas 5.31 and 5.32 you may calculate that Maxwell equation

(5.37) $$\left(\frac{\partial T}{\partial p}\right)_S = \left(\frac{\partial V}{\partial S}\right)_p$$

comes true.

Free variables p and V

(5.38) $$S = \alpha nk \ln\left\{\frac{p\alpha V^{\frac{\alpha+1}{\alpha}}}{\kappa}\right\}$$

(5.39) $$T = \frac{pV}{nk}$$

(5.40) $$U = \alpha pV$$

(5.41) $$H = (\alpha + 1)pV$$

(5.42) $$F = \alpha pV(1 - \ln\left\{\frac{p\alpha V^{\frac{\alpha+1}{\alpha}}}{\kappa}\right\})$$

(5.43) $$G = \alpha pV(1 + \frac{1}{\alpha} - \ln\left\{\frac{p\alpha V^{\frac{\alpha+1}{\alpha}}}{\kappa}\right\})$$

Free variables p and T

(5.44) $$S = \alpha nk \ln\left\{\frac{p\alpha\left(\frac{nkT}{p}\right)^{\frac{\alpha+1}{\alpha}}}{\kappa}\right\}$$

(5.45) $$V = \frac{nkT}{p}$$

(5.46) $$U = \alpha nkT$$

(5.47) $$H = (\alpha + 1)nkT$$

(5.48) $$F = \alpha nkT\left(1 - \ln\left\{\frac{p\alpha\left(\frac{nkT}{p}\right)^{\frac{\alpha+1}{\alpha}}}{\kappa}\right\}\right)$$

(5.49) $$G = \alpha nkT\left(1 + \frac{1}{\alpha} - \ln\left\{\frac{p\alpha\left(\frac{nkT}{p}\right)^{\frac{\alpha+1}{\alpha}}}{\kappa}\right\}\right)$$

With the help of formulas 5.44 and 5.45 you may calculate that Maxwell equation

(5.50) $$-\left(\frac{\partial S}{\partial P}\right)_T = -\left(\frac{\partial V}{\partial T}\right)_p$$

comes true.

Free variables V and T

(5.51) $$S = \alpha nk \ln\left\{\frac{nkTV^{\frac{1}{\alpha}}}{\kappa}\right\}$$

(5.52) $$p = \frac{nkT}{V}$$

(5.53) $$U = \alpha nkT$$

(5.54) $$H = (\alpha + 1)nkT$$

(5.55) $$F = \alpha nkT\left(1 - \ln\left\{\frac{nkTV^{\frac{1}{\alpha}}}{\kappa}\right\}\right)$$

(5.56) $$G = \alpha pV\left(1 + \frac{1}{\alpha} - \ln\left\{\frac{nkTV^{\frac{1}{\alpha}}}{\kappa}\right\}\right)$$

With the help of formulas 5.51 and 5.52 you may calculate that Maxwell equation

(5.57)
$$\left(\frac{\partial S}{\partial V}\right)_T = \left(\frac{\partial p}{\partial T}\right)_V$$

comes true.

# 6. IDEAL GAS THERMODYNAMIC PROCESSES

## 6.1 Introduction

In a thermodynamic process the inner energy of a system is changed by heat transfer and by volume change. In principle there are four variables that change in an ideal gas thermodynamic process – pressure (p), volume (V), temperature (T) and transferred heat energy (Q). The ideal gas inner energy per molecule is the same as temperature. You cannot calculate the end state of a thermodynamic process by straightforwardly using the formulas of the former chapter, because you have to know the amount of the transferred heat and the amount of volume change as functions of time. This is so because the compression energy depends of the state of the gas as explained in chapter 2 'The basic problem' and the state depends on the amount of heat energy transferred to the gas. In a thermodynamic process two quantities are changed freely as functions of time and the values of the other two quantities are calculated. When calculating a process numerically, one proceeds using little time steps (this is the only possible way even for an ideal gas process). Usually one calculates one time step in two steps. In each step one of the quantities is allowed to change while the other one is kept constant. The accuracy of the calculation may be improved by making the used time step smaller. One proceeds in this way calculating the state of the gas step by step by using small time steps until the wanted time value is reached. In this book other kind of a method is used. In case of an ideal gas you are able to calculate the state of the gas analytically, when the two free quantities are changing linearly simultaneously. FORTRAN program which uses this algorithm is presented in appendix 2. Because four quantities define a thermodynamic process, two free variables may be chosen in six different ways. Below necessary formulas are derived for each case, an example input is given for the program and corresponding result is presented as curves.

## 6.2 FORTRAN program

In appendix 2 there is a list of a FORTRAN program that calculates the thermodynamic state of an ideal gas as a function of time, when two quantities are given as functions of time. These two quantities are called free variables. In principle four quantities – pressure (p), volume (V), temperature (T) and transferred heat energy (Q) – govern a thermodynamic process for an ideal gas. Ideal gas inner energy per molecule is the same as temperature. In each calculation case the values of two quantities are given as functions of time and the values for the other two quantities are calculated as functions of time. In addition values for mechanical work and gas entropy are calculated as functions of time. Two quantities out of four may be chosen in six different ways, so that the number of different calculation cases is six. In calculation case 1 the pressure of the ideal gas and the transferred heat energy are given as functions of time. In calculation case 2 the volume of the ideal gas and the transferred heat energy are given as functions of time. In calculation case 3 the temperature of the ideal gas and the transferred heat energy are given as functions of time. In calculation case 4 the pressure and volume of the ideal gas are given as functions of time. In calculation case 5 the pressure and temperature of the ideal gas are given as functions of time. In calculation case 6 the volume and temperature of the ideal gas are given as functions of time.

You may change the calculation case as a function of time and in this way you may calculate different cycle processes. Below you can see the input file of the program for Carnot cycle process. In the first row there are given mission time, number of outputs, number of inner time steps between outputs, the molecular heat capacity of the ideal gas, initial value for pressure, initial value for volume and initial value for temperature. The calculation case as function of time is given in the next three lines. First the number of transition points is given (8 in this case). In the next line the time values for the transition points are given and in the next line the corresponding number (either 1, 2 3, 4, 5 or 6) for the calculation case is given. In this case calculation is done according to calculation case 3 for the first 100 seconds that is temperature and heat transfer are given as functions of time. During 100 – 200 s calculation is done according to calculation case 2 that is volume and heat transfer are given as functions of time. During 200 – 300 s calculations are again done according to calculation case 3 that is temperature and heat transfer are given as functions of time. During the last period 300 – 400 s calculation is again done according to calculation case 2 that is volume and heat transfer are given as functions of time. Pressure change rate as a function of time is given in the next three lines again in 8 time points. In this calculation anything may be given for the pressure change rate, because pressure is not given in any cycle of Carnot process, it is calculated in all cycles. The volume change rate as a function of time is given in the next three lines again in 8 time points. Then the temperature change rate as a function of time is given in the next three lines again in 8 time points. The heat change rate (power) as a function of time is given in the last three lines again in 8 time points. In time period 100 – 200 s volume change rate is given the value 2E-2 m$^3$/s and in time period 300 - 400 s it is given value -0,66556E-2 m3/s. In this case the volume change rate is constant during the whole time periods, but it would be possible to let it follow an arbitrary curve by giving values for it in several time points. The volume change rate (value 2E-2 m$^3$/s) is given also for time period 200 – 300 s, but the program does not use this value, because during this period calculation is done according to calculation case 3 that is temperature and heat transfer are given as functions of time. Next the temperature change rate is given as a function of time in 8 time points. This value is used during time periods 0 – 100 s (value 0 C/s that is temperature is kept constant) and 200 - 300 s (value 0 C/s that is temperature is kept constant). The power to the gas is given in the last three lines. During time period 0 – 100 s power is given value 1,1E3 W, during time period 100 – 200 s power is given value 0 W, during time period 200 – 300 s power is given value -0.897E3 W and during time period 300 – 400 s power is given value 0 W. Carnot process is explained in more detail later, here the program input was introduced using Carnot process as an example.

```
* Ideal gas input data, Carnot process
*
* mission time (s), nbm of prints, inner steps, ALF, P (bar), V (m3), T (C)
400.0Q0 100 1000 2.5Q0 1.0Q0, 1.0Q0, 100.0Q0
* calc mode:1 (P and Q), 2 (V and Q), 3 (T and Q), 4 (P and V), 5 (P and T), 6 (V and T)
8
0.0Q0     100.Q0    100.Q0    200.Q0    200.Q0    300.Q0    300.Q0    400.Q0
3.0       3.0       2.0       2.0       3.0       3.0       2.0       2.0
PRESSURE CHANGE RATE BAR/S AS A FUNKTION OF TIME
8
```

```
0.0Q0    100.Q0   100.Q0   200.Q0   200.Q0   300.Q0   300.Q0   400.Q0
0.0Q-3   0.0Q-3   0.0Q-3   0.0Q-3   0.0Q-3   0.0Q-3   0.0Q-3   0.0Q-3
VOLUME CHANGE RATE M3/S AS A FUNKTION OF TIME
8
0.0Q0    100.Q0   100.Q0   200.Q0   200.Q0   300.Q0   300.Q0        400.Q0
0.0Q-3   0.0Q-3   2.0Q-2   2.0Q-2   2.0Q-2   2.0Q-2   -0.66556Q-2   -0.665560Q-2
TEMPERATURE CHANGE RATE C/S AS A FUNKTION OF TIME
8
0.0Q0    100.Q0   100.Q0   200.Q0   200.Q0   300.Q0   300.Q0   400.Q0
0.0Q0    0.0Q0    0.0Q0    0.0Q0    0.0Q0    0.0Q0    0.0Q0    0.0Q0
THERMAL POWER J/S TO GAS AS A FUNKTION OF TIME
8
0.0Q0    100.Q0   100.Q0   200.Q0   200.Q0      300.Q0      300.Q0   400.Q0
1.1Q3    1.1Q3    0.0Q4    0.0Q4    -0.897Q3    -0.897Q3    0.0Q4    0.0Q4
```

## 6.3 Pressure (p) and volume (V) are given as functions of time

We assume linear dependence during the time step ($\Delta t$)

(6.3.1) $$p = p_0 + at\ ,\ a = \frac{\Delta p}{\Delta t}$$

(6.3.2) $$V = V_0 + bt\ ,\ b = \frac{\Delta V}{\Delta t}$$

Ideal gas obeys always the ideal gas equation, so that temperature as a function of time has value

(6.3.3) $$T = \frac{pV}{nk}$$

and inner energy has value

(6.3.4) $$U = \alpha pV$$

The inner energy changes in this process as follows (we symbolize the inner energy here with letter X instead of U to clarify the explanation)

(6.3.5) $$dX = -pdV = -(p_0 + at)bdt$$

Integrating we get

(6.3.6)

$$X - X_0 = \int_0^{\Delta t} -(p_0 + at)bdt = -(p_0\Delta t + \frac{1}{2}a(\Delta t)^2)\frac{\Delta V}{\Delta t} = -(p_0 + \frac{1}{2}\Delta p)\Delta V$$

where $X_0 = U_0$.

Instead of X the inner energy should have value U so we have to transfer the following amount heat energy to the gas

(6.3.7) $$\Delta Q = U - X$$

Integral for the added heat energy gets value

(6.3.8) $$Q = Q_0 + \Delta Q$$

Now we have calculated values for all quantities (p, V, T, U, Q) at the end of the time step.

An example input for the FORTRAN program is presented below. In the input pressure and volume are given as functions of time. In the example process pressure is decreased linearly during time period 0 – 100 s from value 1 bar to value 0.5 bar, during time period 100 – 200 s pressure is constant 0.5 bar and is increased linearly to value 1 bar during time period 200 -300 s. The other quantity that is given as a function of time is the volume (V) of the gas. Volume is kept constant at value 1 $m^3$ during time period 0 – 150 s and during time period 150 – 300 s the volume of the gas is increased linearly to value 1.5 $m^3$. Temperature (T), heat transfer (Q), mechanical work (W) and entropy (S) are calculated. Temperature is calculated using the ideal gas equation. Heat transfer is determined using the inner energy change caused by the process. Mechanical work is calculated using the defining equation ΔW = -pΔV. Entropy S is calculated using formula 5.38. Constant κ is set to value 1 with no attempt to estimate value for it. Constant κ affects only the absolute level of entropy, so the value for it is not significant with relation to this work. So entropy is calculated using formula (with value 1 for constant κ)

$$S = \alpha nk \ln \left\{ \frac{p\alpha V^{\frac{\alpha+1}{\alpha}}}{\kappa} \right\} \tag{6.3.9}$$

and it does not change, if heat transfer is zero, which is true for this process during time period 100-150 s.

```
* Ideal gas input data, P and V given
*
* mission time (s), nbm of prints, inner steps, ALF, P (bar), V (m3), T (C)
300.0Q0 100 10 2.5Q0 1.0Q0, 1.0Q0, 100.0Q0
* calc mode:1 (P and Q), 2 (V and Q), 3 (T and Q), 4 (P and V), 5 (P and T), 6 (V and T)
6
0.0Q0     100.Q0     100.Q0     200.Q0     200.Q0    300.Q0
4.0       4.0        4.0        4.0        4.0       4.0
*PRESSURE CHANGE RATE BAR/S AS A FUNKTION OF TIME
6
0.0Q0      100.Q0     100.Q0     200.Q0     200.Q0    300.Q0
-0.50Q-2   -0.50Q-2   -0.0Q-2    -0.0Q-2    0.50Q-2   0.50Q-2
*VOLUME CHANGE RATE M3/S AS A FUNKTION OF TIME
4
0.0Q0     150.Q0     150.Q0        300.Q0
0Q2       0Q2        0.333333Q-2   0.333333Q-2
*TEMPERATURE CHANGE RATE C/S AS A FUNKTION OF TIME
6
0.0Q0      100.Q0     100.Q0     200.Q0     200.Q0    300.Q0
-0.0Q0     -0. 0Q0    0.0Q0      0.0Q0      0.0Q0     0.0Q0
*THERMAL POWER J/S TO GAS AS A FUNKTION OF TIME
4
0.0Q0     150.Q0     150.Q0     300.Q0
0Q2       0Q2        0Q2        0Q2
```

The result of the calculation is presented below as curves. To be able to show the curves in same figure the values are normalized so that the maximum of the absolute value is = 1. The real range of values is presented in the legend at the right side of the figure. In addition 'too big' values have been cut off (over 1 or under -1) from efficiency values before drawing the curves. This is done, because efficiency has absolute value infinity, if the volume of the gas changes without heat transfer. The infinite value is prevented by giving initial value 1 J for the transferred heat Q.

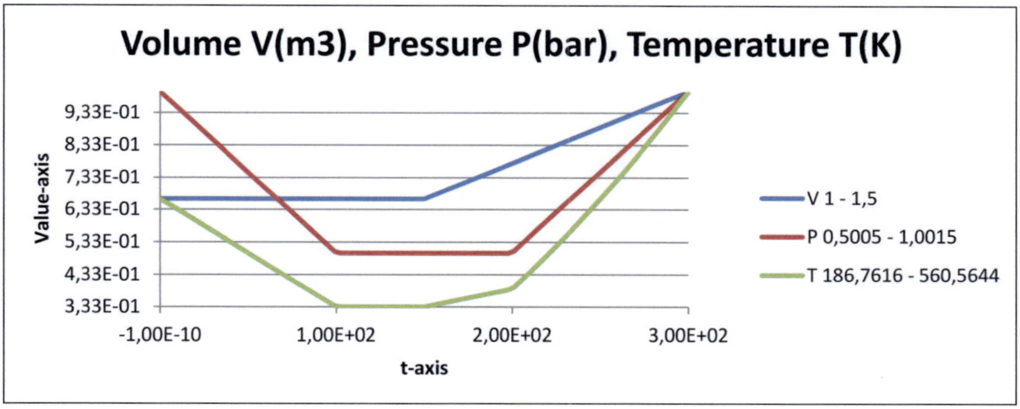


*Figura 6.3.1*

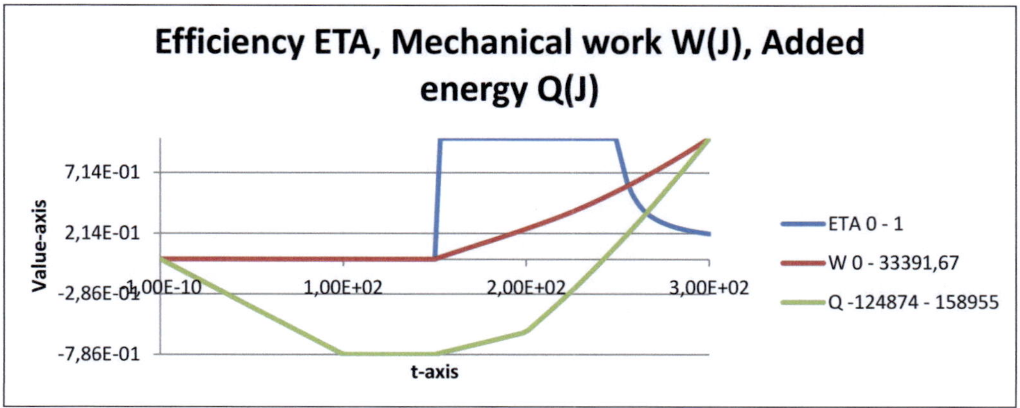


*Figura 6.3.2*

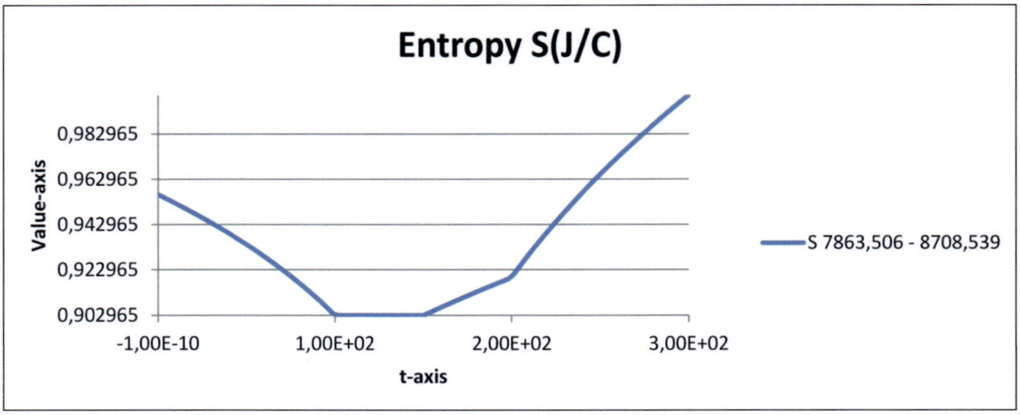


*Figura 6.3.3*

- time period 0-100 s
    - Pressure is decreased linearly from value 1 bar to value 0.5 bars and the volume is kept constant. Because gas obeys the ideal gas equation, temperature decreases. This kind of behavior is possible only, if heat energy is exported from the system. Mechanical energy remains zero, because volume does not change. Entropy decreases, because heat energy is transferred out of the system.
- time period 100 -150 s
    - Both pressure and volume are kept constant, so that also other quantities have constant values.
- time period 150 -200 s
    - Pressure is kept constant and volume is increased linearly from value 1 $m^3$ to value 1.17 $m^3$. Because gas obeys ideal gas equation, temperature increases. Mechanical energy increases, because volume increases. Mechanical energy is subtracted from inner energy, so that temperature should decrease, but it increases. This is possible only if heat energy is imported to the system by heat transfer. Entropy increases, because energy is imported to the system by heat transfer.
- time period 200 -300 s
    - Pressure is increased linearly from value 0.5 bars to value 1.0 bar and volume is increased linearly from value 1.17 $m^3$ to value 1.5 $m^3$ . Because gas obeys ideal gas equation, also temperature increases. Because volume increases, mechanical energy increases and inner energy should decrease with the same amount so that

temperature would decrease. But temperature increases, so that energy must be imported to the system by heat transfer to enable this kind of a physical situation. Entropy increases, because energy is imported to the system by heat transfer.

## 6.4 Pressure (p) and temperature (T) are given as functions of time

We assume linear dependence during the time step ($\Delta t$)

(6.4.1) $$p = p_0 + at\,,\ a = \frac{\Delta p}{\Delta t}$$

(6.4.2) $$T = T_0 + bt\,,\ b = \frac{\Delta T}{\Delta t}$$

Ideal gas obeys always ideal gas equation, so as a function of time volume gets the value

(6.4.3) $$V = \frac{nkT}{p} = nk\frac{T_0+bt}{p_0+at}$$

That is

(6.4.4) $$dV = nk\frac{bp_0-aT_0}{(p_0+at)^2}\,dt$$

Inner energy gets value

(6.4.5) $$U = \alpha nkT$$

The inner energy changes in this process as follows (we symbolize the inner energy here with letter X instead of U to clarify the explanation)

(6.4.6) $$dX = -pdV = -(p_0 + at)nk\frac{bp_0-aT_0}{(p_0+at)^2}\,dt$$

Integrating we get

(6.4.7) $$X - X_0 = \int_0^{\Delta t} -nk\frac{bp_0 - aT_0}{p_0 + at}\,dt = -\,nk\left(\frac{\Delta T p_0}{\Delta p} - T_0\right)\ln\left(1 + \frac{\Delta p}{p_0}\right)$$

$$= nkT_0\ln\left(1 + \frac{\Delta p}{p_0}\right) - nk\Delta T\left(1 - \frac{1}{2}\frac{\Delta p}{p_0} + \frac{1}{3}\left(\frac{\Delta p}{p_0}\right)^2 - \frac{1}{4}\left(\frac{\Delta p}{p_0}\right)^3 + \cdots\right)$$

where $X_0 = U_0$. The logarithmical function has been expressed as a series expansion, because otherwise the formula would diverge in a computer calculation, if pressure is kept constant ($\Delta p = 0$)

Instead of X the inner energy should have value U, so we have to transfer the following amount of heat energy to the gas

(6.4.8) $$\Delta Q = U - X$$

the integral of the added energy is

(6.4.9) $$Q = Q_0 + \Delta Q$$

Now we have calculated values for all quantities (p, V, T, U, Q) at the end of the time step.

An example input for the FORTRAN program is presented below. In the input pressure and temperature are given as functions of time. In this process pressure is decreased linearly during time period 0 – 100 s from value 1 bar to value 0.5 bars, and then kept constant at value 0.5 bar during time period 100 – 200 s and then increased linearly to value 1 bar during time period 200 -300 s. The other quantity that is given as a function of time is the temperature (T) of the gas. Temperature is kept constant at value 100 °C during time period 0 – 150 s and during time period 150 – 300 s the temperature of the gas is increased linearly to value 150 °C. Volume (V), heat transfer (Q), mechanical work (W) and entropy (S) are calculated. Volume is calculated using the ideal gas equation. Heat transfer is determined using the inner energy change caused by the process. Mechanical work is calculated using the defining equation $\Delta W = -p\Delta V$. Entropy S is calculated using formula 5.38. Constant $\kappa$ is set to value 1 with no attempt to estimate value for it. Constant $\kappa$ affects only the absolute level of entropy, so the value for it is not significant with relation to this work. So entropy is calculated using formula (with value 1 for constant $\kappa$)

(6.4.10) $$S = \alpha nk \ln\left\{\frac{p\alpha V^{\frac{\alpha+1}{\alpha}}}{\kappa}\right\}$$

and it does not change, if heat transfer is zero, which is true for this process during time period 100-150 s.

```
* Ideal gas input data, P and T given
*
* mission time (s), nbm of prints, inner steps, ALF, P (bar), V (m3), T (C)
300.0Q0 100 10 2.5Q0 1.0Q0, 1.0Q0, 100.0Q0
* calc mode:1 (P and Q), 2 (V and Q), 3 (T and Q), 4 (P and V), 5 (P and T), 6 (V and T)
6
0.0Q0     100.Q0     100.Q0     200.Q0     200.Q0     300.Q0
5.0       5.0        5.0        5.0        5.0        5.0
*PRESSURE CHANGE RATE BAR/S AS A FUNKTION OF TIME
6
0.0Q0      100.Q0      100.Q0     200.Q0     200.Q0     300.Q0
-0.50Q-2   -0.50Q-2    -0.0Q-2    -0.0Q-2    0.50Q-2    0.50Q-2
*VOLUME CHANGE RATE M3/S AS A FUNKTION OF TIME
4
0.0Q0     150.Q0     150.Q0     300.Q0
0Q2       0Q2        0.0Q-2     0.0Q-2
*TEMPERATURE CHANGE RATE C/S AS A FUNKTION OF TIME
4
0.0Q0     150.Q0     150.Q0         300.Q0
0Q2       0Q2        0.333330Q0     0.333330Q0
*THERMAL POWER J/S TO GAS AS A FUNKTION OF TIME
4
0.0Q0     150.Q0     150.Q0     300.Q0
0Q2       0Q2        0Q2        0Q2
```

The result of the calculation is presented below as curves. To be able to show the curves in same figure the values are normalized so that the maximum of the absolute value is = 1. The real range of values is presented in the legend at the right side of the figure.

In addition 'too big' values have been cut off (over 1 or under -1) from efficiency values before drawing the curves. This is done, because efficiency has absolute value infinity, if the volume of the gas changes without heat transfer.

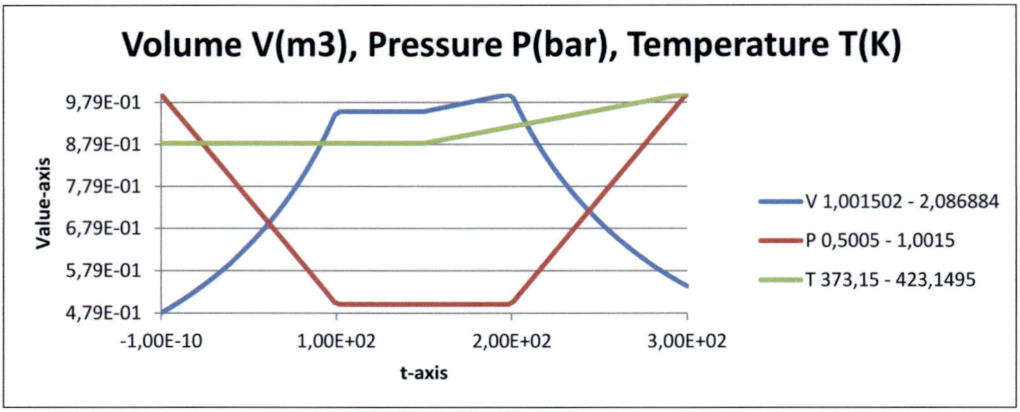


*Figure 6.4.1*

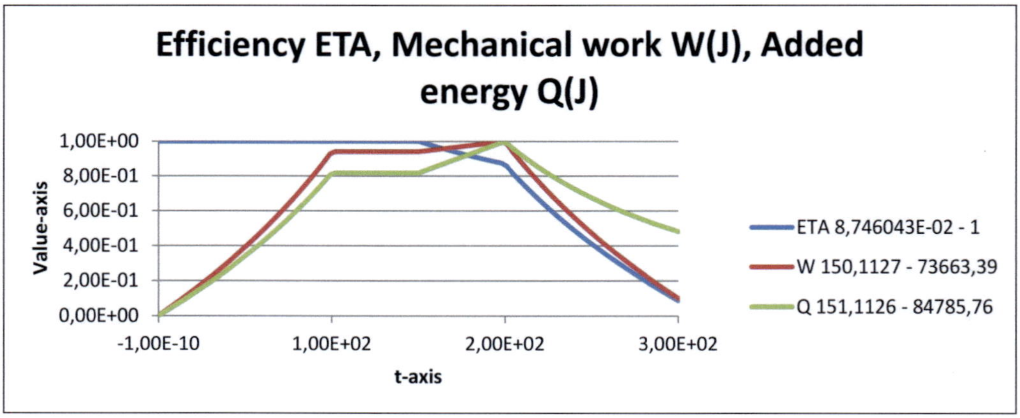


*Figure 6.4.2*

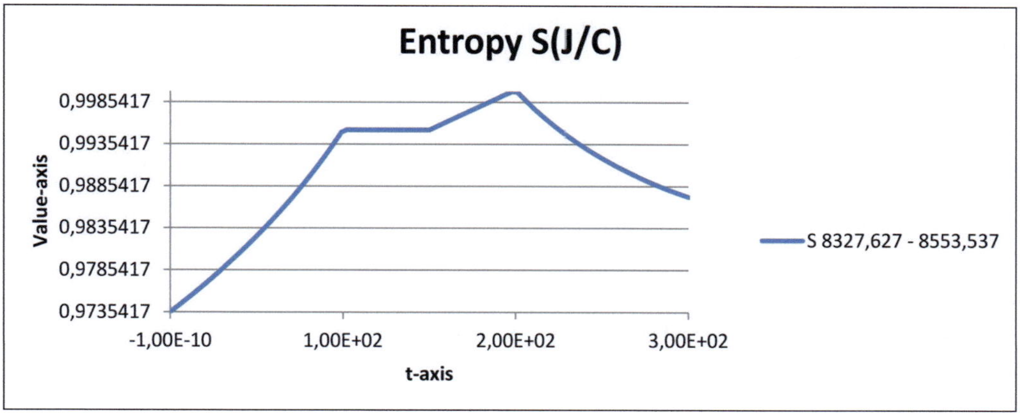


*Figure 6.4.3*

- time period 0-100 s
  - Pressure is decreased linearly from value 1 bar to value 0.5 bars and the temperature is kept constant. Because gas obeys the ideal gas equation, volume of the gas must increase. This would cause the decrease of inner energy and temperature. But temperature is constant so heat energy must be imported to the system. Mechanical energy increases, because volume increases. Entropy increases, because heat energy is transferred to the system.
- time period 100 -150 s
  - Both pressure and temperature are kept constant, so that also other quantities have constant values.
- time period 150 -200 s
  - Pressure is kept constant and temperature is increased linearly from value 100 °C to value 117 °C. Because gas obeys ideal gas equation, volume must increase. Mechanical energy increases, because volume increases. Mechanical energy is subtracted from inner energy, so that temperature should decrease, but it increases. This is possible only if heat energy is imported to the system by heat transfer. Entropy increases, because energy is imported to the system by heat transfer.
- time period 200 -300 s
  - Pressure is increased linearly from value 0.5 bars to value 1.0 bar and temperature is increased linearly from value 117 °C to value 150 °C. Volume is calculated with the help of ideal gas equation, and it decreases with these changes in pressure and

temperature. Because volume decreases, mechanical energy decreases and inner energy should increase with the same amount so that temperature would increase. But the temperature increase due to mechanical work would be too big, so that energy must be exported from the system by heat transfer. Entropy decreases, because energy is exported from the system by heat transfer.

## 6.5 Volume (V) and temperature (T) are given as functions of time

We assume linear dependence during the time step ($\Delta t$)

(6.5.1) $$V = V_0 + at\,, a = \frac{\Delta V}{\Delta t}$$

(6.5.2) $$T = T_0 + bt\,, b = \frac{\Delta T}{\Delta t}$$

So that

(6.5.3) $$dV = adt$$

Ideal gas obeys always ideal gas equation, so that pressure gets as a function of time value

(6.5.4) $$p = \frac{nkT}{V} = nk\frac{T_0+bt}{V_0+at}$$

Inner energy gets value

(6.5.5) $$U = nkT$$

The inner energy changes in this process as follows (we symbolize the inner energy here with letter X instead of U to clarify the explanation)

(6.5.6) $$dX = -pdV = -nk\frac{T_0+bt}{V_0+at}a\,dt = -nka(\frac{T_0-\frac{b}{a}V_0}{V_0+at} + \frac{b}{a}t)dt$$

By integration we get

(6.5.7)

$$X - X_0 = \int_0^{\Delta t} -nka\left(\frac{T_0 - \frac{b}{a}V_0}{V_0 + at} + \frac{b}{a}t\right) dt = -nk\left(\left(T_0 - \frac{\Delta T}{\Delta V}V_0\right)\ln\left(1 + \frac{\Delta V}{V_0}\right) + \Delta T\right)$$

$$= -nk\left(T_0 \ln\left(1 + \frac{\Delta V}{V_0}\right) + \Delta T\right) + nk\Delta T\left(1 - \frac{1}{2}\frac{\Delta V}{V_0} + \frac{1}{3}\left(\frac{\Delta V}{V_0}\right)^2 - \frac{1}{4}\left(\frac{\Delta V}{V_0}\right)^3 + \cdots\right)$$

where $X_0 = U_0$. The logarithmic function has been expressed as a series expansion, because otherwise the formula would diverge in a computer calculation, if volume is kept constant ($\Delta V = 0$).

Instead of X the inner energy should have value U so we have to transfer the following amount of heat energy to the gas

(6.5.8)
$$\Delta Q = U - X$$

and the integrated value for the added energy is

(6.5.9)
$$Q = Q_0 + \Delta Q$$

Now we have calculated values for all quantities (p, V, T, U, Q) at the end of the time step.

An example input for the FORTRAN program is presented below. In the input volume and temperature are given as functions of time. In this process volume is decreases linearly during time period 0 – 100 s from value 1 $m^3$ to value 0.5 $m^3$, and then kept constant at value 0.5 $m^3$ during time period 100 – 200 s and then increased linearly to value 1 $m^3$ during time period 200 -300 s. The other quantity that is given as a function of time is the temperature (T) of the gas. Temperature is kept constant at value 100 °C during time period 0 – 150 s and during time period 150 – 300 s the temperature of the gas is increased linearly to value 150 °C. Pressure (p), heat transfer (Q), mechanical work (W) and entropy (S) are calculated. Pressure is calculated using the ideal gas equation. Heat transfer is determined using the inner energy change caused by the process. Mechanical work is calculated using the defining equation $\Delta W = -p\Delta V$. Entropy S is calculated using formula 5.38. Constant $\kappa$ is set to value 1 with no attemp to estimate value for it. Constant $\kappa$ affects only the absolute level of entropy, so the value for it is not significant with relation to this work. So entropy is calculated using formula (with value 1 for constant $\kappa$)

$$S = \alpha nk \ln \left\{ \frac{p\alpha V^{\frac{\alpha+1}{\alpha}}}{\kappa} \right\} \tag{6.5.10}$$

and it does not change, if heat transfer is zero, which is true for this process during time period 100-150 s.

```
* Ideal gas input data, V and T given
*
* mission time (s), nbm of prints, inner steps, ALF, P (bar), V (m3), T (C)
300.0Q0 100 10 2.5Q0 1.0Q0, 1.0Q0, 100.0Q0
* calc mode:1 (P and Q), 2 (V and Q), 3 (T and Q), 4 (P and V), 5 (P and T), 6 (V and T)
6
0.0Q0     100.Q0    100.Q0    200.Q0    200.Q0   300.Q0
6.0       6.0       6.0       6.0       6.0      6.0
*PRESSURE CHANGE RATE BAR/S AS A FUNKTION OF TIME
6
0.0Q0     100.Q0    100.Q0    200.Q0    200.Q0   300.Q0
-0.0Q-2   -0.0Q-2   -0.0Q-2   -0.0Q-2   0.0Q-2   0.0Q-2
*VOLUME CHANGE RATE M3/S AS A FUNKTION OF TIME
6
0.0Q0     100.Q0    100.Q0    200.Q0    200.Q0   300.Q0
-0.5Q-2   -0.5Q-2   -0.0Q-2   -0.0Q-2   0.5Q-2   0.5Q-2
*TEMPERATURE CHANGE RATE C/S AS A FUNKTION OF TIME
4
0.0Q0     150.Q0    150.Q0         300.Q0
0Q2       0Q2       0.333330Q0     0.333330Q0
*THERMAL POWER J/S TO GAS AS A FUNKTION OF TIME
4
0.0Q0     150.Q0    150.Q0    300.Q0
0Q2       0Q2       0Q2       0Q2
```

The result of the calculation is presented below as curves. To be able to show the curves in same figure the values are normalized so that the maximum of the absolute value is = 1. The real range of values is presented in the legend at the right side of the figure. In addition 'too big' values have been cut off (over 1 or under -1) from efficiency values before drawing the curves. This is done, because efficiency has absolute value infinity, if the volume of the gas changes without heat transfer.

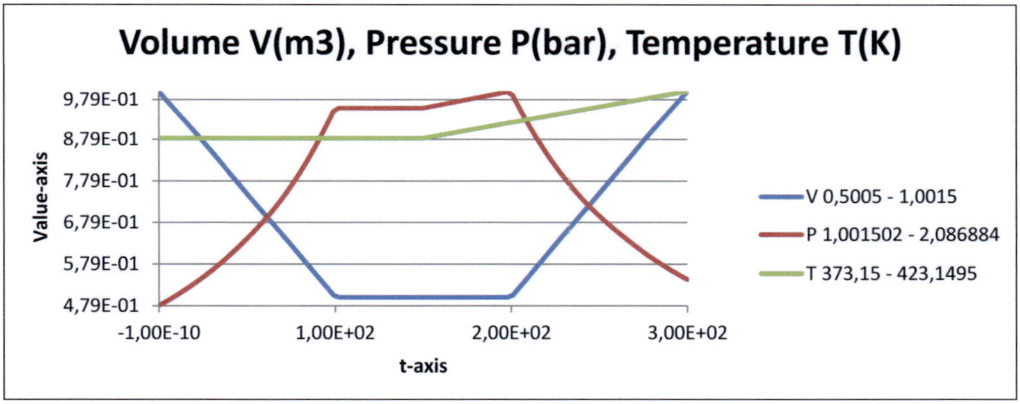


*Figure 6.5.1*

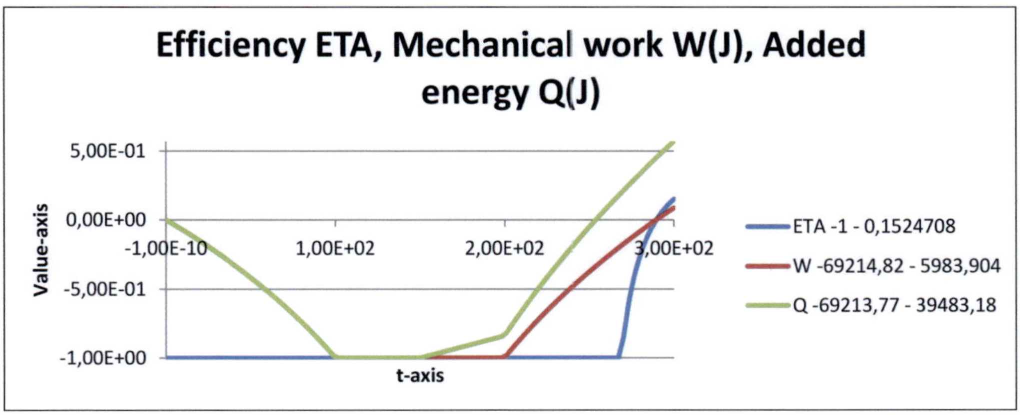


*Figure 6.5.2*

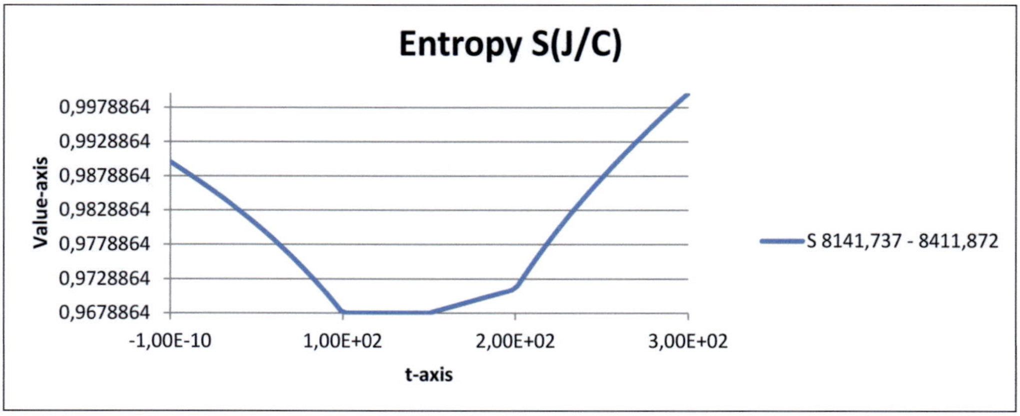


*Figure 6.5.3*

- time period 0-100 s
  - Volume is decreased linearly from value 1 $m^3$ to value 0.5 $m^3$ and the temperature is kept constant. Because gas obeys the ideal gas equation, pressure of the gas must increase. Because volume decreases mechanical work should increase the gas inner energy and temperature of the gas. Temperature however is kept constant so that heat energy must be exported from the gas. Mechanical energy and heat transfer are equal in value so that they are overlapping in figure 6.5.2. Entropy decreases, because heat energy is transferred out of the system.
- time period 100 -150 s
  - Both volume and temperature are kept constant, so that also other quantities have constant values.
- time period 150 -200 s
  - Volume is kept constant and temperature is increased linearly from value 100 °C to value 117 °C. Because gas obeys ideal gas equation, pressure must increase. Inner energy increases linearly proportional to the temperature, so that heat energy must be added to the gas by heat transfer. Mechanical energy is constant, because volume is constant. Entropy increases, because energy is imported to the system by heat transfer.
- time period 200 -300 s
  - Volume is increased linearly from value 0.5 $m^3$ to value 1.0 $m^3$ and temperature is increased linearly from value 117 °C to value 150 °C. Pressure is calculated with the

help of ideal gas equation. Increase of temperature increases pressure and increase of volume decreases pressure. The increase of volume is proportionally greater so that pressure decreases. Because volume increases, mechanical energy increases so that inner energy and temperature should decrease. But the temperature increases so heat energy must be added to the system. Entropy increases, because energy is imported to the system through heat transfer.

## 6.6 Volume (V) and heat transfer (Q) are given as functions of time

We assume linear dependence during the time step ($\Delta t$)

(6.6.1)
$$V = V_0 + at\,,\; a = \frac{\Delta V}{\Delta t}$$

(6.6.2)
$$Q = Q_0 + bt\,,\; b = \frac{\Delta Q}{\Delta t}$$

In this process inner energy changes as follows

(6.6.3)
$$dU = -pdV + bdt = -\frac{U}{\alpha(V_0+at)}adt + bdt$$

Next we change variable t to y

(6.6.4)
$$y = \ln(V_0 + at)\,,\; dy = \frac{adt}{V_0+at}\,,\; dt = \frac{V_0+at}{a}dy$$

We get

(6.6.5)
$$\frac{dU}{dy} = -\frac{U}{\alpha} + \frac{b}{a}e^{y}$$

Insert a trial function

(6.6.6)
$$U = Be^{-\frac{y}{\alpha}} + Ae^{y}$$

and you get the following value for A

(6.6.7) $$A = \frac{\alpha}{(\alpha+1)}\frac{b}{a}$$

So that

(6.6.8) $$U = Be^{-\frac{y}{\alpha}} + \frac{\alpha}{(\alpha+1)}\frac{b}{a}e^{y} = B\left(\frac{1}{V_0+at}\right)^{\frac{1}{\alpha}} + \frac{\alpha}{(\alpha+1)}\frac{b}{a}(V_0 + at)$$

Use the initial condition U = $U_0$, V = $V_0$ , when t = 0 and you get

(6.6.9) $$B = V_0^{\frac{1}{\alpha}}\left(U_0 - \frac{\alpha}{\alpha+1}\frac{b}{a}V_0\right)$$

That is

(6.6.10) $$U = V_0^{\frac{1}{\alpha}}\left(U_0 - \frac{\alpha}{\alpha+1}\frac{b}{a}V_0\right)\left(\frac{1}{V_0+at}\right)^{\frac{1}{\alpha}} + \frac{\alpha}{(\alpha+1)}\frac{b}{a}(V_0 + at)$$

At the end of the time step

(6.6.11) $$U = \left(U_0 - \frac{\alpha}{\alpha+1}\frac{\Delta Q}{\Delta V}V_0\right)\left(\frac{V_0}{V_0+\Delta V}\right)^{\frac{1}{\alpha}} + \frac{\alpha}{(\alpha+1)}\frac{\Delta Q}{\Delta V}(V_0 + \Delta V)$$

This formula diverges in a computer calculation, if the volume is kept constant ($\Delta V = 0$). To avoid divergence the exponent term is expressed as a series expansion

(6.6.12) $$U = U_0\left(\frac{V_0}{V_0+\Delta V}\right)^{\frac{1}{\alpha}} + \frac{\alpha}{\alpha+1}\Delta Q\left(1 + \frac{\frac{1}{\alpha}+\frac{1-\alpha}{2\alpha^2}\frac{\Delta V}{V_0}+\frac{(1-\alpha)(1-2\alpha)}{6\alpha^3}\left(\frac{\Delta V}{V_0}\right)^2+}{1+\frac{\Delta V}{V_0}\left(\frac{1}{\alpha}+\frac{1-\alpha}{2\alpha^2}\frac{\Delta V}{V_0}+\frac{(1-\alpha)(1-2\alpha)}{6\alpha^3}\left(\frac{\Delta V}{V_0}\right)^2+\right)}\right)$$

The following formula

(6.6.13) $$(1 + x)^{\frac{i}{j}} = 1 + \frac{i}{j}x + \frac{i(i-j)}{j2j}x^2 + \frac{i(i-j)(i-2j)}{j2j3j}x^3 + \cdots$$

is used to construct the series expansion of the exponential function. Temperature is calculated using the equation

(6.6.14) $$T = \frac{U}{\alpha nk}$$

Pressure is calculated using the ideal gas equation

(6.6.15) $$p = \frac{nkT}{V}$$

Now we have calculated values for all quantities (p, V, T, U, Q) at the end of the time step.

An example input for the FORTRAN program is presented below. In the input volume and heat transfer are given as functions of time. In this process volume is decreased linearly during time period 0 – 100 s from value 1 $m^3$ to value 0.5 $m^3$, and then kept constant at value 0.5 $m^3$ during time period 100 – 200 s and then increased linearly to value 1 $m^3$ during time period 200 -300 s. The other quantity that is given as a function of time is the heat transfer (Q) to the gas. Heat transfer is zero during time period 0 – 150 s and during time period 150 – 300 s total amount of 120 kJ heat is added with constant power to the gas. Pressure (p), temperature (T), mechanical work (W) and entropy (S) are calculated. Temperature is calculated using the inner energy change (formulas 6.6.12 and 6.6.14) and pressure is calculated using the ideal gas equation. Mechanical work is calculated using the defining equation $\Delta W = -p\Delta V$. Entropy S is calculated using formula 5.38. Constant $\kappa$ is set to value 1 with no attempt to estimate value for it. Constant $\kappa$ affects only the absolute level of entropy, so the value for it is not significant with relation to this work. So entropy is calculated using formula (with value 1 for constant $\kappa$)

(6.6.16) $$S = \alpha nk \ln\left\{\frac{p\alpha V^{\frac{\alpha+1}{\alpha}}}{\kappa}\right\}$$

and it does not change (although p and V change so as the process demands), if heat transfer is zero, which is true for this process during time period 0-150 s.

```
* Ideal gas input data, V and Q given
*
* mission time (s), nbm of prints, inner steps, ALF, P (bar), V (m3), T (C)
300.0Q0 100 10 2.5Q0 1.0Q0, 1.0Q0, 100.0Q0
* calc mode:1 (P and Q), 2 (V and Q), 3 (T and Q), 4 (P and V), 5 (P and T), 6 (V and T)
```

```
6
0.0Q0     100.Q0     100.Q0     200.Q0     200.Q0    300.Q0
2.0       2.0        2.0        2.0        2.0       2.0
*PRESSURE CHANGE RATE BAR/S AS A FUNKTION OF TIME

6
0.0Q0     100.Q0     100.Q0     200.Q0     200.Q0    300.Q0
-0.0Q-2   -0.Q-2     -0.0Q-2    -0.0Q-2    0.0Q-2    0.0Q-2
*VOLUME CHANGE RATE M3/S AS A FUNKTION OF TIME
6
0.0Q0      100.Q0     100.Q0    200.Q0    200.Q0     300.Q0
-0.50Q-2  -0.50Q-2   0.0Q-2     0.0Q-2    0.50Q-2   0.50Q-2
*TEMPERATURE CHANGE RATE C/S AS A FUNKTION OF TIME
6
0.0Q0     100.Q0     100.Q0     200.Q0     200.Q0    300.Q0
0.0Q0     0.0Q0      0.0Q0      0.0Q0      0.0Q0     0.0Q0
*THERMAL POWER J/S TO GAS AS A FUNKTION OF TIME
4
0.0Q0     150.Q0     150.Q0    300.Q0
0Q2       0Q2        8Q2       8Q2
```

The result of the calculation is presented below as curves. To be able to show the curves in same figure the values are normalized so that the maximum of the absolute value is = 1. The real range of values is presented in the legend at the right side of the figure. In addition 'too big' values have been cut off (over 1 or under -1) from efficiency values before drawing the curves. This is done, because efficiency has absolute value infinity, if the volume of the gas changes without heat transfer.

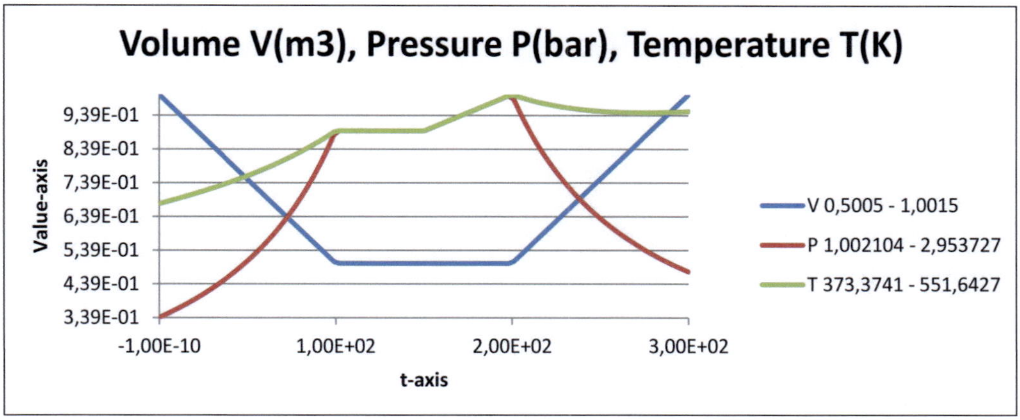


*Figure 6.6.1*

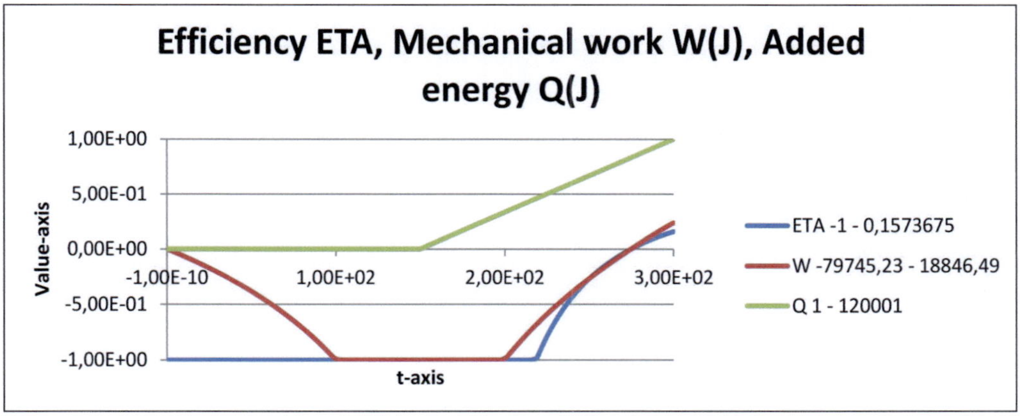


*Figure 6.6.2*

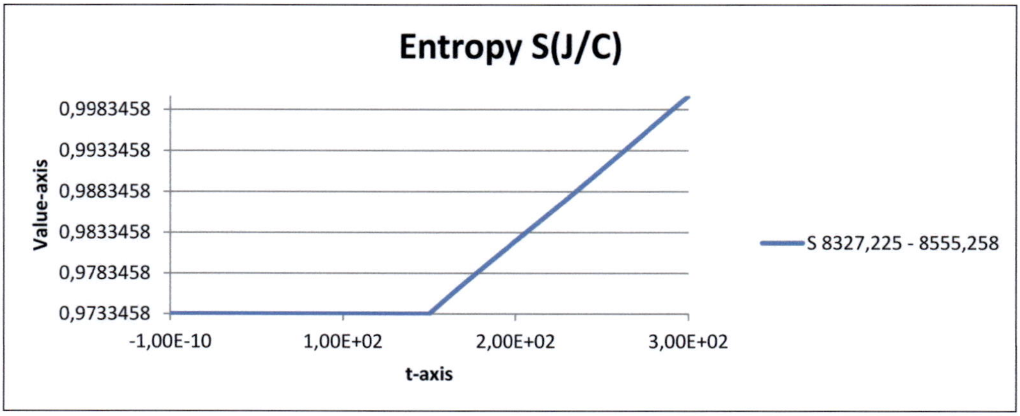


*Figure 6.6.3*

- time period 0-100 s
  - Volume is decreased linearly from value 1 $m^3$ to value 0.5 $m^3$ and heat transfer is zero. Because volume decreases, mechanical work increases the inner energy and temperature of the gas. Pressure is given by the ideal gas equation and it increases, because volume decreases and temperature increases. Entropy is constant, because heat transfer is zero.
- time period 100 -150 s
  - Volume is kept constant and heat transfer is zero so that also other quantities have constant values.
- time period 150 -200 s
  - Volume is kept constant and a total amount of 40 kJ of heat energy is transferred linearly to the gas. Inner energy and temperature increase. Pressure is got from the ideal gas equation and it increases. Mechanical energy is constant, because volume is constant. Entropy increases, because energy is imported to the system by heat transfer.
- time period 200 -300 s
  - Volume is increased linearly from value 0.5 $m^3$ to value 1.0 $m^3$ and a total amount of 80 kJ of heat energy is transferred linearly to the gas. Because volume increases, mechanical work increases. Mechanical work decreases inner energy and heat transfer increases inner energy. Mechanical work is greater, so that temperature decreases and pressure decreases according to ideal gas equation Entropy increases, because energy is imported to the system through heat transfer.

## 6.7 Pressure (p) and heat transfer (Q) are given as functions of time

We assume linear dependence during the time step ($\Delta t$)

(6.7.1) $$p = p_0 + at\,, a = \frac{\Delta p}{\Delta t}$$

(6.7.2) $$Q = Q_0 + bt\,, b = \frac{\Delta Q}{\Delta t}$$

In this process inner energy changes in the following way

(6.7.3) $$dU = -pdV + bdt = -(p_0 + at)dV + bdt$$

(6.7.4) $$V = \frac{nkT}{p} = \frac{U}{\alpha p}$$

(6.7.5) $$\frac{dV}{dt} = \frac{\frac{dU}{dt}}{\alpha p} - \frac{U}{\alpha p^2}a$$

So that

(6.7.6) $$\frac{dU}{dt} = -p\left(\frac{\frac{dU}{dt}}{\alpha p} - \frac{U}{\alpha p^2}a\right) + b$$

Or equivalently

(6.7.7) $$\frac{dU}{dt} = \frac{aU}{\alpha+1}\frac{1}{p_0+at} + \frac{\alpha}{\alpha+1}b$$

Next we perform a variable change

(6.7.8) $$y = \ln(p_0 + at)\,,\ dy = \frac{adt}{p_0+at}$$

We get

(6.7.9) $$\frac{d(aU)}{dy} = \frac{aU}{\alpha+1} + \frac{\alpha}{\alpha+1}be^y$$

Insert the following trial function

(6.7.10) $$aU = Be^{\frac{y}{\alpha+1}} + Ae^{y}$$

and A gets the value

(6.7.11) $$A = b$$

so

(6.7.12) $$aU = B(p_0 + at)^{\frac{1}{\alpha+1}} + b(p_0 + at)$$

Use the initial condition $U = U_0$, $p = p_0$, when $t = 0$ and you will get

(6.7.13) $$B = \frac{aU_0 - bp_0}{p_0^{\frac{1}{\alpha+1}}}$$

So that

(6.7.14) $$U = U_0\left(\frac{p_0+at}{p_0}\right)^{\frac{1}{\alpha+1}} + \Delta Q\frac{p_0}{\Delta p}\left(1 - \left(\frac{p_0+at}{p_0}\right)^{\frac{1}{\alpha+1}}\right) + \Delta Q$$

At the end of the time step you get

(6.7.15) $$U = U_0\left(1 + \frac{\Delta p}{p_0}\right)^{\frac{1}{\alpha+1}} + \Delta Q + \Delta Q\frac{p_0}{\Delta p}\left(1 - \left(1 + \frac{\Delta p}{p_0}\right)^{\frac{1}{\alpha+1}}\right)$$

This formula diverges in a computer calculation, if the pressure is kept constant ($\Delta p = 0$). To avoid the divergence the exponent term is expressed as a series expansion

(6.7.16) $$U = U_0\left(\frac{p_0+\Delta p}{p_0}\right)^{\frac{1}{\alpha+1}} + \Delta Q(1 - \frac{1}{\alpha+1} - \frac{-\alpha}{2(\alpha+1)^2}\frac{\Delta p}{p_0} - \frac{-\alpha(-2\alpha-1)}{6(\alpha+1)^3}\left(\frac{\Delta p}{p_0}\right)^2 - \cdots)$$

The following formula

(6.7.17) $$(1+x)^{\frac{i}{j}} = 1 + \frac{i}{j}x + \frac{i(i-j)}{j2j}x^2 + \frac{i(i-j)(i-2j)}{j2j3j}x^3 + \cdots$$

is used to construct the series expansion of the exponent function. Temperature is calculated using the equation

$$T = \frac{U}{\alpha n k} \tag{6.7.18}$$

Volume is calculated using the ideal gas equation

$$V = \frac{nkT}{p} \tag{6.7.19}$$

Now we have calculated values for all quantities (p, V, T, U, Q) at the end of the time step.

An example input for the FORTRAN program is presented below. In the input pressure and heat transfer are given as functions of time. In this process pressure is decreased linearly during time period 0 – 100 s from value 1 bar to value 0.5 bars, and then kept constant at value 0.5 bar during time period 100 – 200 s and then increased linearly to value 1 bar during time period 200 -300 s. The other quantity that is given as a function of time is the heat transfer (Q) to the gas. Heat transfer is zero during time period 0 – 150 s and during time period 150 – 300 s a total amount of 120 kJ of heat is added with a constant power to the gas. Volume (V), temperature (T), mechanical work (W) and entropy (S) are calculated. Temperature is calculated using the inner energy change (formulas 6.7.16 and 6.7.18) and volume is calculated using the ideal gas equation. Mechanical work is calculated using the defining equation $\Delta W = -p\Delta V$. Entropy S is calculated using formula 5.38. Constant κ is set to value 1 with no attempt to estimate value for it. Constant κ affects only the absolute level of entropy, so the value for it is not significant with relation to this work. So entropy is calculated using formula (with value 1 for constant κ)

$$S = \alpha n k \ln\left\{\frac{p\alpha V^{\frac{\alpha+1}{\alpha}}}{\kappa}\right\} \tag{6.7.20}$$

and it does not change (although p and V change so as the process demands), if heat transfer is zero, which is true for this process during time period 0-150 s.

```
* Ideal gas input data, P and Q given
*
* mission time (s), nbm of prints, inner steps, ALF, P (bar), V (m3), T (C)
300.0Q0 100 10 2.5Q0 1.0Q0, 1.0Q0, 100.0Q0
* calc mode:1 (P and Q), 2 (V and Q), 3 (T and Q), 4 (P and V), 5 (P and T), 6 (V and T)
6
0.0Q0    100.Q0    100.Q0    200.Q0    200.Q0   300.Q0
1.0      1.0       1.0       1.0       1.0      1.0
*PRESSURE CHANGE RATE BAR/S AS A FUNKTION OF TIME
6
0.0Q0     100.Q0     100.Q0    200.Q0    200.Q0   300.Q0
-0.50Q-2  -0.50Q-2   0.0Q-2    0.0Q-2    0.50Q-2  0.50Q-2
*VOLUME CHANGE RATE M3/S AS A FUNKTION OF TIME
6
0.0Q0     100.Q0    100.Q0    200.Q0    200.Q0   300.Q0
-0.0Q-2   -0.Q-2    -0.0Q-2   -0.0Q-2   0.0Q-2   0.0Q-2
*TEMPERATURE CHANGE RATE C/S AS A FUNKTION OF TIME
6
0.0Q0    100.Q0    100.Q0    200.Q0    200.Q0   300.Q0
0.0Q0    0.0Q0     0.0Q0     0.0Q0     0.0Q0    0.0Q0
*THERMAL POWER J/S TO GAS AS A FUNKTION OF TIME
4
0.0Q0    150.Q0    150.Q0   300.Q0
0Q2      0Q2       8Q2      8Q2
```

The result of the calculation is presented below as curves. To be able to show the curves in same figure the values are normalized so that the maximum of the absolute value is = 1. The real range of values is presented in the legend at the right side of the figure. In addition 'too big' values have been cut off (over 1 or under -1) from efficiency values before drawing the curves. This is done, because efficiency has absolute value infinity, if the volume of the gas changes without heat transfer.

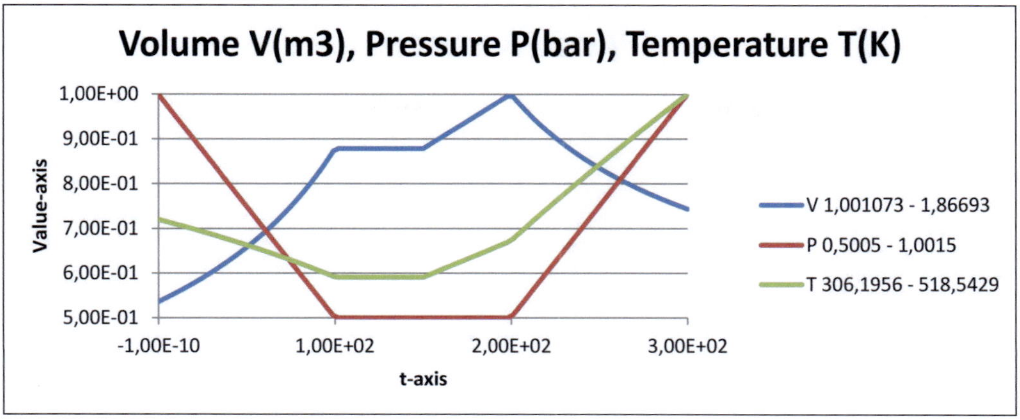


*Figure 6.7.1*

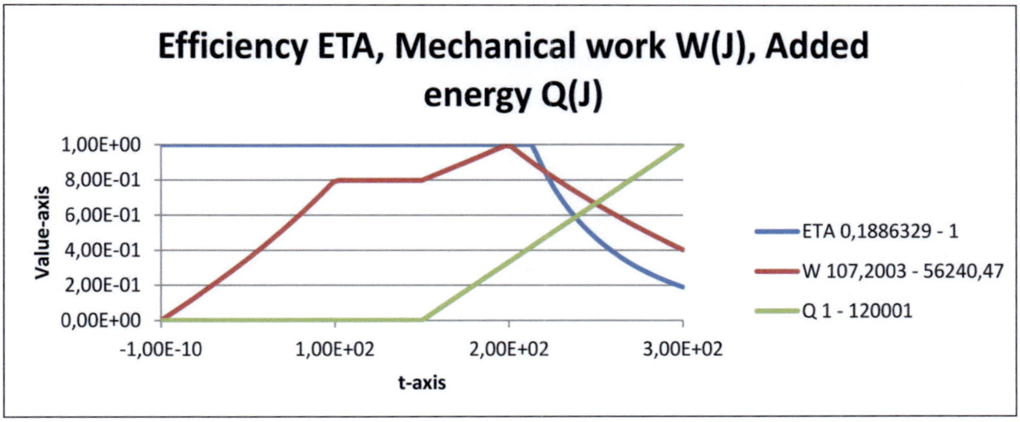


*Figure 6.7.2*

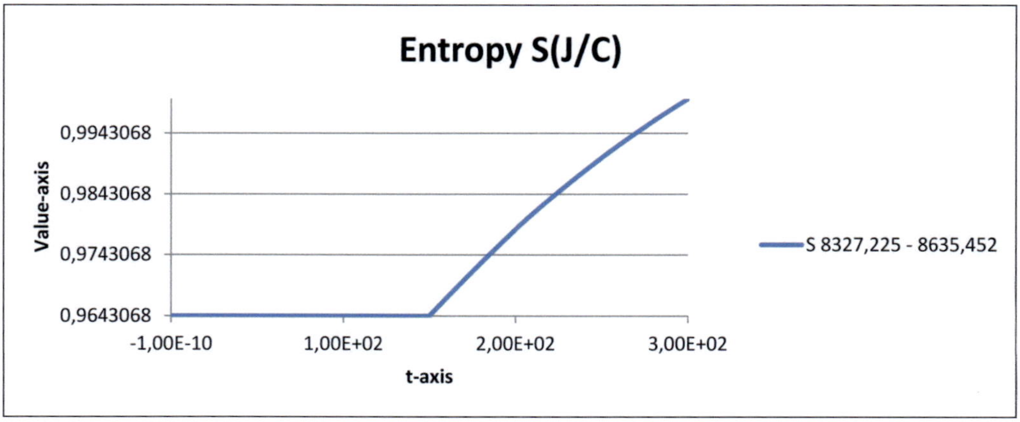


*Figure 6.7.3*

- time period 0-100 s
  - Pressure is decreased linearly from value 1 bar to value 0.5 bars and heat transfer is zero. Because volume increases, mechanical work increases and the inner energy and temperature of the gas decrease. Entropy is constant, because heat transfer is zero.
- time period 100 -150 s
  - Pressure is kept constant and heat transfer is zero so that also other quantities have constant values.
- time period 150 -200 s
  - Pressure is kept constant and 40 kJ of heat energy is transferred linearly to the gas. Volume increases and gas makes mechanical work, but decrease of inner energy due to mechanical work is less than the heat energy imported by heat transfer so that temperature increases. Entropy increases, because energy is imported to the system by heat transfer.
- time period 200 -300 s
  - Pressure is increased linearly from value 0.5 bar to value 1.0 bar and 80 kJ of heat energy is transferred linearly to the gas. Inner energy is calculated as a function of pressure increment and increment of inner energy due to heat transfer and temperature may be calculated using the inner energy and it increases. Volume may be calculated using the ideal gas equation and it decreases so that mechanical work is done to the gas. Entropy increases, because energy is imported to the system through heat transfer.

## 6.8 Temperature (T) and heat transfer (Q) are given as functions of time

We assume linear dependence during the time step ($\Delta t$)

(6.8.1) $$T = T_0 + at\,, a = \frac{\Delta T}{\Delta t}$$

(6.8.2) $$Q = Q_0 + bt\,, b = \frac{\Delta Q}{\Delta t}$$

According to ideal gas equation

(6.8.3) $$p = \frac{nk(T_0+at)}{V}$$

In this process inner energy changes in the following way

(6.8.4) $$U = \alpha nkT$$

So that

(6.8.5) $$dU = \alpha nkadt$$

We get

(6.8.6) $$dU = \alpha nkadt = -pdV + bdt = -\frac{nkT}{V}dV + bdt$$

so that

(6.8.7) $$(\alpha nka - b)dt = -\frac{nk(T_0+at)}{V}dV$$

By integrating we get at the end of the time step ($t = \Delta t$)

(6.8.8) $$V = (1 + \frac{\Delta T}{T_0})^{\frac{\Delta Q}{nk\Delta T}-\alpha}V_0$$

This formula diverges in a computer calculation, if the temperature is kept constant ($\Delta t = 0$). To avoid the divergence the exponent term is expressed as a series expansion

(6.8.9)

$$V = V_0(1 + \frac{\Delta T}{T_0})^{-\alpha}\left(1 + \frac{\Delta Q}{nk}\frac{1}{T_0} + \frac{\Delta Q(\Delta Q - nk\Delta T)}{2(nk)^2}\left(\frac{1}{T_0}\right)^2 + \frac{\Delta Q(\Delta Q - nk\Delta T)(\Delta Q - 2nk\Delta T)}{6(nk)^3}\left(\frac{1}{T_0}\right)^3 + \cdots\right)$$

The following formula

(6.8.10) $$(1 + x)^{\frac{i}{j}} = 1 + \frac{i}{j}x + \frac{i(i-j)}{j2j}x^2 + \frac{i(i-j)(i-2j)}{j2j3j}x^3 + \cdots$$

is used to construct the series expansion for the exponent function. Pressure is calculated using the ideal gas equation

(6.8.11) $$p = \frac{nkT}{V}$$

Now we have calculated values for all quantities (p, V, T, U, Q) at the end of the time step.

An example input for the FORTRAN program is presented below. In the input temperature and heat transfer are given as functions of time. In this process temperature is decreased linearly during time period 0 – 100 s from value 100 °C to value 50 °C , and then kept constant at value 50 °C during time period 100 – 200 s and then increased linearly to value 100 °C during time period 200 -300 s. The other quantity that is given as a function of time is the heat transfer (Q) to the gas. Heat transfer is zero during time period 0 – 150 s and during time period 150 – 300 s a total amount of 120 kJ of heat is added with constant power to the gas. Volume (V), pressure (T), mechanical work (W) and entropy (S) are calculated. Volume is defined by the inner energy change (formulas 6.8.6 and 6.8.8) and pressure is calculated using the ideal gas equation. Mechanical work is calculated using the defining equation ΔW = -pΔV. Entropy S is calculated using formula 5.38. Constant κ is set to value 1 with no attempt to estimate value for it. Constant κ affects only the absolute level of entropy, so the value for it is not significant with relation to this work. So entropy is calculated using formula (with value 1 for constant κ)

(6.8.12)

$$S = \alpha nk \ln\left\{\frac{p\alpha V^{\frac{\alpha+1}{\alpha}}}{\kappa}\right\}$$

and it does not change (although p and V change so as the process demands) if heat transfer is zero, which is true for this process during time period 0-150 s.

```
* Ideal gas input data, T and Q given
*
* mission time (s), nbm of prints, inner steps, ALF, P (bar), V (m3), T (C)
300.0Q0 100 10 2.5Q0 1.0Q0, 1.0Q0, 100.0Q0
* calc mode:1 (P and Q), 2 (V and Q), 3 (T and Q), 4 (P and V), 5 (P and T), 6 (V and T)
6
0.0Q0    100.Q0    100.Q0    200.Q0    200.Q0    300.Q0
3.0      3.0       3.0       3.0       3.0       3.0
*PRESSURE CHANGE RATE BAR/S AS A FUNKTION OF TIME
6
0.0Q0    100.Q0    100.Q0    200.Q0    200.Q0    300.Q0
-0.0Q-2  -0.Q-2    -0.0Q-2   -0.0Q-2   0.0Q-2    0.0Q-2
*VOLUME CHANGE RATE M3/S AS A FUNKTION OF TIME
6
0.0Q0    100.Q0    100.Q0    200.Q0    200.Q0    300.Q0
-0.0Q-2  -0.0Q-2   0.0Q-2    0.0Q-2    0.0Q-2    0.0Q-2
*TEMPERATURE CHANGE RATE C/S AS A FUNKTION OF TIME
6
0.0Q0    100.Q0    100.Q0    200.Q0    200.Q0    300.Q0
-0.50Q0  -0.50Q0   0.0Q0     0.0Q0     0.50Q0    0.50Q0
*THERMAL POWER J/S TO GAS AS A FUNKTION OF TIME
4
0.0Q0    150.Q0    150.Q0    300.Q0
0Q2      0Q2       8Q2       8Q2
```

The result of the calculation is presented below as curves. To be able to show the curves in same figure the values are normalized so that the maximum of the absolute value is = 1. The real range of values is presented in the legend at the right side of the figure. In addition 'too big' values have been cut off (over 1 or under -1) from efficiency values before drawing the curves. This is done, because efficiency has absolute value infinity, if the volume of the gas changes without heat transfer.

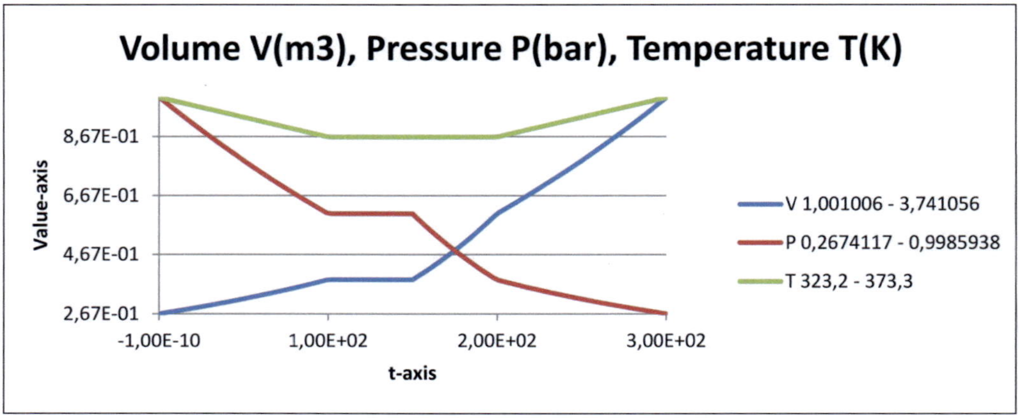


*Figure 6.8.1*

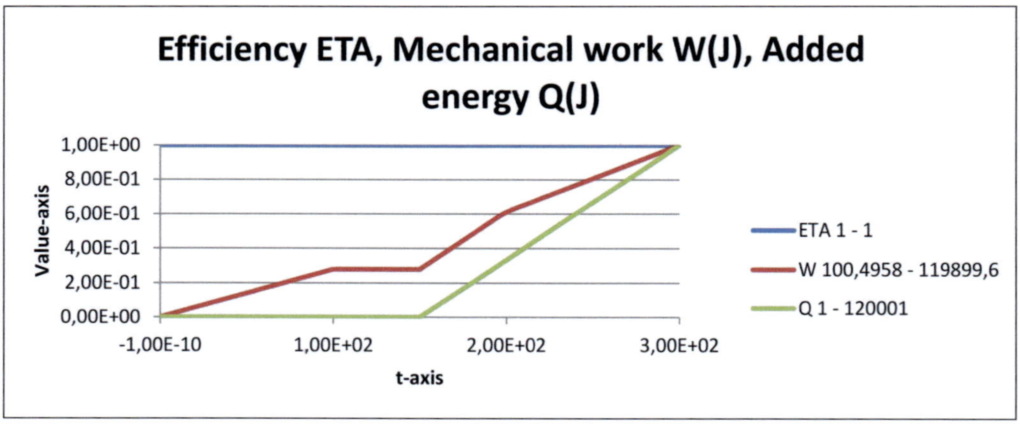


*Figure 6.8.2*

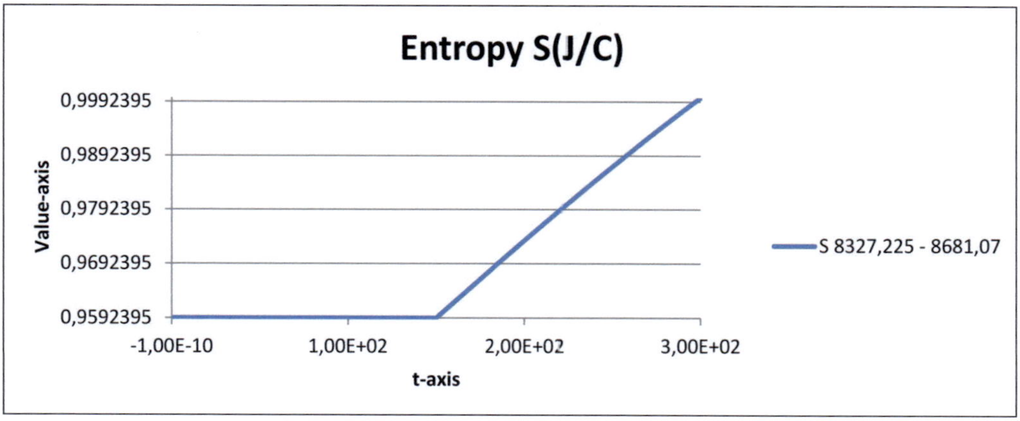


*Figure 6.8.3*

- time period 0-100 s
  - Temperature is decreased linearly from value 100 °C to value 50 °C and heat transfer is zero. The inner energy of the gas decreases proportionally to temperature. Because heat transfer is zero, the only way to decrease inner energy is the mechanical work so that volume increases. Volume is calculated with help of inner energy change and pressure is calculated using ideal gas equation. Entropy is constant, because heat transfer is zero.
- time period 100 -150 s
  - Temperature is kept constant and heat transfer is zero so that also other quantities have constant values.
- time period 150 -200 s
  - Temperature is kept constant and 40 kJ of heat energy is transferred linearly to the gas. Because temperature is constant, the imported heat energy must be exported as mechanical work so that volume increases. Pressure is calculated using the ideal gas equation and it decreases. Entropy increases, because energy is imported to the system by heat transfer.
- time period 200 -300 s
  - Temperature is increased linearly from value 50 °C to value 100 °C and 80 kJ of heat energy is transferred linearly to the gas. 80 kJ energy addition would increase the temperature more than 50 °C so that the extra energy must be exported from the gas as mechanical work so that the volume increases. Pressure is calculated

using the ideal gas equation and it decreases. Entropy increases, because energy is imported to the system through heat transfer.

## 6.9 Carnot process

The Carnot process is an ideal cyclic process that is impossible to realize in practice. Carnot process produces maximal amount of mechanical energy out of thermal energy. Carnot process is composed of four cycles. After fourth cycle the state of the 'machine' is the same as the initial state. Carnot process that uses ideal gas consists of the following four cycles.

1. Thermal energy is imported to the gas in such a way that the temperature of the gas stays constant. This is possible only if the gas expands in such a way that all the imported heat energy transforms into mechanical energy.
2. Gas expands without heat transfer. In this cycle the pressure and the temperature of the gas decrease.
3. Heat energy is exported from the gas in such a way that the temperature of the gas stays constant. This is possible only, if gas is compressed simultaneously in such a way, that the amount of mechanical work is equal to the exported heat energy.
4. Gas is compressed without heat transfer. In this cycle the pressure and temperature of the gas are increased.

Because there is no heat transfer in cycles 2 and 4, entropy stays constant in these cycles. Theoretically you may import whatever amount of energy in cycle 1 to the gas. In cycle 2 you may theoretically allow the gas to expand to any volume you like. In cycle 3 you have to export such an amount of energy out of the gas that entropy decreases to the value it had at the beginning of cycle 1. Entropy does not change in cycle 4, so that after cycle 4 entropy has its initial value. In cycle four the volume of the gas is decreased to the initial value (to the value it had at the beginning of cycle 1). After these four cycles the combined change in volume and the combined change in entropy is zero, so that also other quantities have their initial values (the values they had at the beginning of cycle 1), because volume change and entropy change define the gas state uniquely.

Input for the FORTRAN program for the four cycles of the Carnot process is presented below.

```
* Ideal gas input data, Carnot process
*
* mission time (s), nbm of prints, inner steps, ALF, P (bar), V (m3), T (C)
400.0Q0 100 1000 2.5Q0 1.0Q0, 1.0Q0, 100.0Q0
* calc mode:1 (P and Q), 2 (V and Q), 3 (T and Q), 4 (P and V), 5 (P and T), 6 (V and T)
8
0.0Q0    100.Q0   100.Q0   200.Q0   200.Q0   300.Q0   300.Q0   400.Q0
3.0      3.0      2.0      2.0      3.0      3.0      2.0      2.0
PRESSURE CHANGE RATE BAR/S AS A FUNKTION OF TIME
8
0.0Q0    100.Q0   100.Q0   200.Q0   200.Q0   300.Q0   300.Q0   400.Q0
0.0Q-3   0.0Q-3   0.0Q-3   0.0Q-3   0.0Q-3   0.0Q-3   0.0Q-3   0.0Q-3
VOLUME CHANGE RATE M3/S AS A FUNKTION OF TIME
8
0.0Q0    100.Q0   100.Q0   200.Q0   200.Q0   300.Q0   300.Q0   400.Q0
0.0Q-3   0.0Q-3   2.0Q-2   2.0Q-2   2.0Q-2   2.0Q-2   -0.66556Q-2 -0.665560Q-2
TEMPERATURE CHANGE RATE C/S AS A FUNKTION OF TIME
8
0.0Q0    100.Q0   100.Q0   200.Q0   200.Q0   300.Q0   300.Q0   400.Q0
0.0Q0    0.0Q0    0.0Q0    0.0Q0    0.0Q0    0.0Q0    0.0Q0    0.0Q0
THERMAL POWER J/S TO GAS AS A FUNKTION OF TIME
8
0.0Q0    100.Q0   100.Q0   200.Q0   200.Q0   300.Q0   300.Q0   400.Q0
1.1Q3    1.1Q3    0.0Q4    0.0Q4    -0.897Q3 -0.897Q3 0.0Q4    0.0Q4
```

The result of the calculation is presented below as curves. To be able to show the curves in same figure the values are normalized so that the maximum of the absolute value is = 1. The real range of values is presented in the legend at the right side of the figure. The efficiency of a cyclic process is defined as the sum of mechanical work of all cycles divided by the amount of the heat energy imported to the gas. The imported heat energy is the energy added to the gas, the heat energy transferred from the gas by heat transfer is not subtracted from the added heat energy.

Another way to get the efficiency of a Carnot process is (as is well known)

(6.9.1) $$ETA = 1 - \frac{T_L}{T_H}$$

You may read in the legend of figure 6.9.1 that $T_L$ = 304,2568 °K and $T_H$ = 373,15 °K, so that you get the value 0.1846 for the efficiency, which is the same value, which the program has calculated from energy balance for the end of cycle 4, as can be seen in figure 6.9.2.

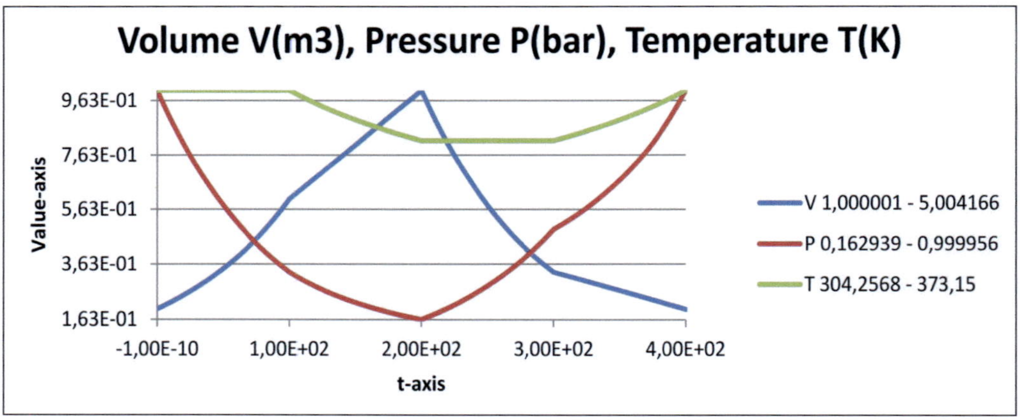


*Figure 6.9.1*

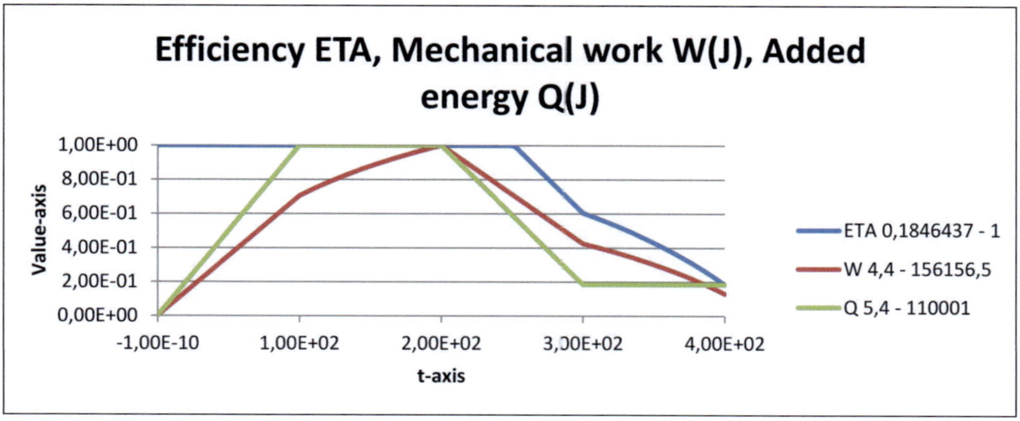


*Figure 6.9.2*

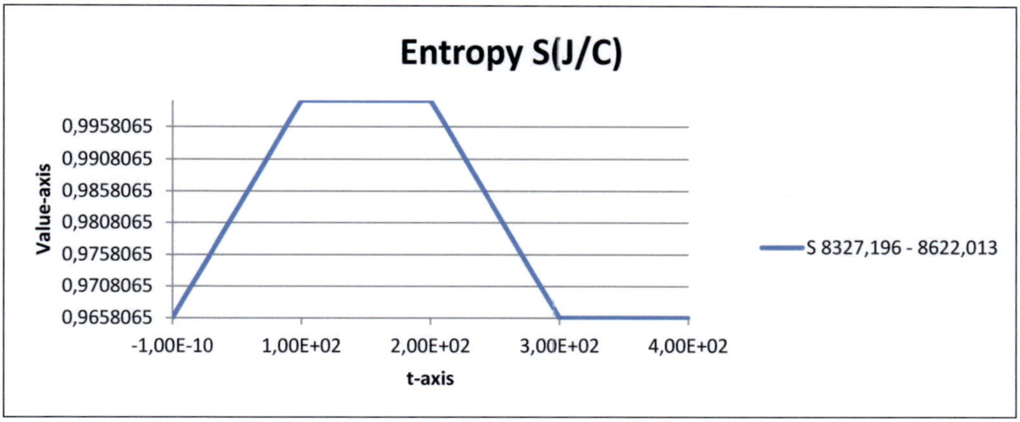


*Figure 6.9.3*

- time period 0 -100 s
  - Temperature is kept constant and heat energy is transferred to the gas with a constant power of 1.1 kJ/s. Because temperature is constant, the imported energy must be transferred by mechanical work out of the gas, so volume increases. Pressure is calculated using the ideal gas equation and it decreases. Entropy increases, because energy is imported to the gas by heat transfer. In this way you may in theory transfer to the gas as much energy as you want to by choosing appropriate value for the power to the gas.

- time period 200 -300 s
    - Volume is increased with a constant speed. You may increase the volume so much as you want to by choosing appropriate value for speed of volume change. Heat transfer is zero. Gas makes mechanical work, because volume increases. Mechanical work is subtracted from the inner energy, so that temperature decreases and pressure decreases according to the ideal gas equation.  Entropy is constant, because heat transfer is zero
- time period 200 -300 s
    - Temperature is kept constant and heat energy is exported from the gas. This is possible only if the gas is compressed in such a way that the mechanical work added to the gas equals the energy that is exported from the gas by heat transfer. Entropy decreases, because energy is exported from the gas by heat transfer. In this cycle such an amount of energy must be exported from the gas that entropy decreases to its initial value (the value at the beginning of cycle 1). This was easy to do by trying different power values for the heat transfer that is by manual iteration of the power value.
- time period 300 -400 s
    - Volume is decreased to its initial value that is to the value it had at the beginning of cycle 1. Because heat transfer is zero, entropy is constant. So that after cycle 4 the volume and the entropy have their initial values that is the values they had at the beginning of cycle 1. Other quantities have also their initial values, because volume and entropy define uniquely the state of the gas.

# 7. REAL MATERIALS

You have to know the dependence of the measurable quantities p,V and T, which corresponds to the ideal gas equation, for example as a numeric table. So you have the relation in one form or other

(7.1) $$p = p(V, T)$$

Additionally you have to know the relation between inner energy and temperature, In case of real materials the relation depends also on pressure and volume so you have the relation

(7.2) $$U = U(T, p, V)$$

in one form or other for example as a numeric table. You may solve T and p using formulas 7.1 and 7.2 (for example numerically) as functions of V and U. To get the inner energy as a function of S and V you may proceed in the same way as in the case of ideal gas. So you insert T(V,U) and p(V,U) into formulas 5.5 and 5.6 and solve the partial differential equations (for example numerically if you don't have analytic expressions). In this way you may calculate all thermodynamic properties of the system starting from equations 7.1 and 7.2.

Thermodynamic processes are calculated in the same way as in the case of ideal gas. If you don't have analytic expressions for the functional dependence of the quantities, you may use numeric tables for the thermodynamic properties for the materials in question.

# APPENDIX 1

## LABOURS OF JORACLES, MATHEMATICAL ALGORITHMS FOR TECHNICAL CALCULATIONS

### LABOUR 1: THERMODYNAMICS OF WATER (1983 – 1984)

- Water, steam, and gases are inserted into a containment building
- The task is to calculate the pressure and temperature of the containment as a function of time
- Before the labour computer programs could compute approximately right in certain situations, when parameters of the program had been chosen appropriately using results of many experiments. The program might create/destroy mass or energy, but an experienced user of the program could however calculate reasonable results by choosing appropriate parameter values suitable for the case under consideration. So that to get an accurate result many expensive experiments were needed and demanding use of programs adjusted by parameters to the case under consideration was required. Using this kind of a method is of course very slow and expensive.
- Labour: Pressure and temperature calculation of water + steam + gas –mixture using mass end energy conservation equations as the starting point. Results are always right and mass and energy is not created/destroyed artificially.
- Consideration of the labour: Thermodynamics was developed at 1800 century, so that there have been over 100 years of time to develop a functional calculation method for the situation described above. Evidently it was considered impossible to develop an exact calculation algorithm for the situation described above, although the theory of thermodynamics was known well.

## LABOUR 2: EXACT NUMERICAL CALCULATION OF THE HEAT CONDUCTION EQUATION (1984)

- Heat conduction equation is the simplest of the so called conservation equations
- Heat conduction equation is a partial differential equation for the temperature field in which second order differentiation is made regarding position and first order differentiation is made regarding time
- Before the labour people tried to improve the numerical calculation results by reducing the so called nodal division inside the body and by reducing the time step. This led to very long calculation times with a computer, and the right result could not be achieved with the numerical methods with any nodal division or time step. This was verified by computing such a case that could be solved also analytically (for example the time development of the temperature field of a homogeneous iron ball that has been sunk into hot oil). These old programs also created or destroyed energy incorrectly, but using parameter values fitted to experimental results one could calculate temperature field time development also in conditions that differ a bit from the conditions of the experiments. So that to get an accurate result many expensive experiments were needed and demanding use of programs adjusted by parameters to the case under consideration was required. Using this kind of a method is of course very slow and expensive.
- The labour: Developing a nodal calculation algorithm using energy conversation as the starting point. Long time steps may be used and sparse spatial nodal division may be used and the result is always right (without any suspicious parameters) as calculation of the time dependent temperature field of for example homogenous iron ball sunk into hot oil proves.
- Consideration of the labour: Heat conduction equation was developed in 1800 –century and computers capable for quick numerical calculation were constructed in 1960's, so that there had been over 20 year's time to construct an algorithm that calculates right the problem described above. Because a numeric algorithm may be used also without a computer (although slowly) time for construction has been almost 100 years. In technology there has been the need for exact calculation of the temperature field at least from 1920-30's. But one could not do it before the labour 2 of Joracles. Apparently the exact calculation of the temperature field was considered impossible, although the theory of heat transfer was known well.

## LABOUR 3: EXACT NUMERICAL CALCULATION OF THE POINT KINETIC EQUATION (1985)

- Integrating the time dependent neutron diffusion equation over the reactor core you get a differential equation the solution of which gives the reactor power as a function of time. The equation is called the point kinetic equation.
- As a matter of fact you get a group of differential equations, because every delayed neutron group must be calculated using a differential equation of its own and in addition the amount of iodine and xenon (strong reactivity poison) must be calculated using their own differential equations.
- The differential equation group is classified as stiff, because it includes time constants that are very different in length (the generation time of sequential neutron generations – about $10^{-4}$ seconds, time constants for delayed neutrons – about 0,01 -100 seconds and half lives of iodine and xenon – about 7 hour)
- You get an approximate result for a stiff differential equation by routine numeric methods (for example Runge-Kutta-method), but you are not even near the exact result.
- After taking an examination of a book that dealt only the solving of the point kinetic equation and after getting a task to construct a program that solves point kinetic equation Joracles got a new idea for solving numerically a differential equation.
- The idea in the old methods is (always?) to go in the direction of the derivative of the solution an amount defined by the time step. But because in the book there were presented many exact solutions for such a bit simplified point kinetic equation, in which some quantity (for example reactor temperature) is kept constant, Joracles got the idea that one could follow the exact (exponential) solution of the simplified equation during the time step and then perhaps calculate a new value for the quantity that was kept constant and perform the time step again using this new value for the quantity (implicit like method).
- When using Runge-Kutta-method the solution may escape from the right track but when using the method of Joracles the right solution track is followed.

## LABOUR 4: EXACT NUMERIC CALCULATION OF NEUTRON DIFFUSION/TRANSPORT EQUATION (1989)

- The reactor power distribution, neutron flux distribution and total power are determined by solving the time dependent neutron diffusion equation for the reactor core
- The numeric solution of neutron diffusion (or transport) equation had a tendency to diverge. Decreasing the time step and making the nodal division denser didn't solve this problem. By adding unphysical 'inertia' to the equations helped a bit, but this kind of a solution is artificial.
- When using the so called implicit method the nodal values of quantities stand for values at the end of the time step in the difference equation that approximates the differential equation in question. This improves numerical stability. This was of course tried also in the case of solving the neutron diffusion equation, but the algorithm behaved still in an unstable way. The mistake that was made was that although the values for neutron flux stood for the end time of the time step the temperature of the uranium fuel was not treated in an implicit way, but the values for it were left unchanged as the values at the beginning of the time step. The increase of fuel temperature controls down the increase of neutron flux, so if the temperature of the fuel is not allowed to change during the time step the solution tends to diverge. The labour number 4 of Joracles was to develop such a numeric algorithm for neutron flux calculation in which the change in fuel temperature is taken into account in an implicit way.
- And the calculations began to go fluently. The solution didn't diverge, not when using infinitely small time steps neither when using shocking long time steps. In addition the solution had such an unexpected and inexperienced characteristic that the result was not depended on the length of the time step (the same miracle as in labours 1 and 2).
- Now even an inexperienced computer user could get right results with rather poor skills and experienced parameter value user was not needed anymore. Former experienced parameter user could get rather good results after choosing appropriate parameter values for his/her calculation algorithms and the appropriate parameters values had been determined by many expensive experiments.

## LABOUR 5: EXACT NUMERICAL CALCULATION OF 1-DIMENSIONAL FLOW EQUATION – NAVIER-STOKES-EQUATION (1990)

- Flow equation (group) is a very unstable equation and its numerical solutions tend to diverge. Chaos theory stands as a proof for this. Chaos theory was born because of this instability of flow equations (very small changes in initial conditions cause quite different solution in due time – butterfly wing beat in Andalusia cause severe twister in India after two months).
- Numeric solution of neutron diffusion equation tends to diverge and the solution of flow equations tends to diverge. AHAA. As a matter of fact at the same seconds as computer calculation gave the result that the solution of neutron diffusion equation doesn't diverge, if you treat the temperature of uranium fuel in an implicit way, I was 90 % sure that the same fault was done also in the numeric solution of flow equations. In this case the quantity that was even in implicit methods not treated in implicit way was pressure. It is the pressure difference that pushes the flow and if the pressure difference is kept in the initial value of the time step, the flow accelerates too much and the solution diverges.
- When pressure is treated in an implicit way the calculation begins to go fluently. You may use so long a time step as you like (with some wide limits of course) and the spatial nodalization doesn't have to be very dense. In addition the result is independent of time step and spatial nodalization; these are properties that the old methods could not reach.

## LABOUR 6: EXACT NUMERICAL CALCULATION OF 2-DIMENSIONAL – ALSO 3 – ALSO MANY DIMENSIONAL NAVIER-STOKES FLOW EQUATION (1990)

- In addition to last paragraph two dimensional impulse transfer from node to node was needed. This was implemented conserving the total impulse and as a result you got an algorithm that produced fine flow patterns. Let's mention that Navier-Stokes equations don't conserve the total impulse, so when taking the differential limit (going from difference equation to differential equation) one notices that one is not actually solving the Navier-Stokes equations. Question arises: which equations actually describe flow properly.

For the end one may say that these labours take the numerical technical calculation to a new level and save enormous amounts of money and lengthen the collective life extend of human beings with a really remarkable amount already now and especially in the future.

# APPENDIX 2

## FORTRAN PROGRAM FOR IDEAL GAS THERMODYNAMIC PROCESS CALCULATION

```
! IdealGas.f90
!
! FUNCTIONS:
! IdealGas - Entry point of console application.
!

!****************************************************************************
!
! PROGRAM: IdealGas
!
! PURPOSE:  Entry point for the console application.
!
!****************************************************************************

    program IdealGas




    implicit none

    ! Variables
   integer(8)  I, J, M, L, MA, II
   integer(8) NAA,NAAS
   real(16) P, V, T, U, W, Q, P0, V0, T0, U0, W0, Q0, PA, VA, TA, UA, WA, QA, S, ETA
   real(16) PMAX, VMAX, TMAX, UMAX, WMAX, QMAX, SMAX, ETAMAX
   real(16) PMIN, VMIN, TMIN, UMIN, WMIN, QMIN, SMIN, ETAMIN
   real(16) PM, VM, TM, UM, WM, QM, SM, ETAM
```

```
real(16) N, K, ALF, KAPS, ETA1, ERRO, R, RR, X
real(16) MAIKA,  AIKA, AA
INTEGER LP, LV, LT, LQ, LCHO, LASK
real(16) SP, SV, ST, SQ, CHO
REAL(16) PZ(30,2), VZ(30,2), TZ(30,2), QZ(30,2), CHOZ(30,2)
real(16) VAPU, JAPU
DATA N /6.02252Q23/ ! AVAGARDOS CONSTANT
DATA K /1.38054Q-23/ ! BOLTZMAN CONSTANT

! Body of Console7

OPEN(UNIT=1,FILE='IgAlkuarvot.txt') ! Progran input file
OPEN(UNIT=2,FILE='IgTulos.txt') ! Output file
OPEN(UNIT=3,FILE='IgPrintAPU.txt') ! Quantities to be drawn as curves
OPEN(UNIT=4,FILE='IgPrint.txt',DECIMAL = 'COMMA') ! Quentities to be
transfered to excel and drawn as curves

CALL LAUSEET(3) ! read 3 (dummy) comment lines in input file
! read mission time, nbr prints, inner steps, heat capacity, P, V, T
READ(1,*) MAIKA, NAA, NAAS, ALF, P0, V0, T0
CALL LAUSEET(1)
! read calculation mode as a function of time
READ(1,*) LCHO
READ(1,*) ((CHOZ(I,J),I=1,LCHO),J=1,2)
CALL LAUSEET(1)
! read pressure change rate as a function of time
READ(1,*) LP
READ(1,*) ((PZ(I,J),I=1,LP),J=1,2)
CALL LAUSEET(1)
! read volume change rate as a function of time
READ(1,*) LV
READ(1,*) ((VZ(I,J),I=1,LV),J=1,2)
CALL LAUSEET(1)
! read temperature change rate as a function of time
```

```
READ(1,*) LT
READ(1,*) ((TZ(I,J),I=1,LT),J=1,2)
CALL LAUSEET(1)
! read power to gas as a function of time
READ(1,*) LQ
READ(1,*) ((QZ(I,J),I=1,LQ),J=1,2)


KAPS = 273.15Q0 ! absolut zero temperature
T0 = T0 + KAPS ! C -> K
P0 = P0*1Q5     ! BAR -> PASCAL
AA = MAIKA/(NAA*NAAS) ! time step
W0 = 0Q0
Q0 = 1.0Q0

T = T0
V = V0
P = P0
W = W0
Q = Q0

N = P*V/(K*T)

U0 = ALF*N*K*T0
U = U0

PA = P
VA = V
TA = T
UA = U
QA = Q
ETA = 1
AIKA = -1.0Q-10
S = ALF*N*K*QLOG((ALF*P*V**((ALF+1)/ALF))/1Q0)
```

```
  VMAX = 0.0Q0
  PMAX = 0.0Q0
  TMAX = 0.0Q0
  UMAX = 0.0Q0
  WMAX = 1.0Q-10
  QMAX = 1.0Q-10
  SMAX = 0.0Q0
  ETAMAX = 0.0Q0
  VMIN = 10.0Q10
  PMIN = 10.0Q10
  TMIN = 10.0Q10
  UMIN = 10.0Q10
  WMIN = 10.0Q10
  QMIN = 10.0Q10
  SMIN = 10.0Q10
  ETAMIN = 10.0Q10
  VM = 1.0Q1
  PM = 1.0Q1
  TM = 1.0Q1
  UM = 1.0Q1
  WM = 1.0Q1
  QM = 1.0Q1
  SM = 1.0Q1
  ETAM = 1.0Q1
!  CALL MINMAX
  WRITE(2,*) ,AIKA,     V,     VAPU,     P,     T,     U,     W,     E,  S, ETA'
  WRITE(2,1)  AIKA, V, VAPU, P/1Q5, T-KAPS, U, W, Q, S, ETA
  WRITE(2,*)
  WRITE(3,2)  AIKA, V, P, T, U, W, Q, S, ETA
  LASK = 1
  DO I = 1, NAA
  DO J = 1, NAAS
    AIKA = AIKA+AA
! get change rates for quantities
```

```
    CALL ARVI(CHO,AIKA,30,LCHO,CHOZ)
    CALL ARVI(SP,AIKA,30,LP,PZ)
    CALL ARVI(SV,AIKA,30,LV,VZ)
    CALL ARVI(ST,AIKA,30,LT,TZ)
    CALL ARVI(SQ,AIKA,30,LQ,QZ)
    SP = SP*1.0Q5
! quantity values at the begining of the time step
    TA = T
    UA = U
    PA = P
    VA = V
    QA = Q
!     WRITE(*,*) AIKA, TA, UA, PA, VA
! pressure and heat flow are given as functions of time in input
     IF(CHO.LT.1.5) THEN
       P = PA + SP*AA   ! pressure
       Q = QA + SQ*AA ! integral of added heat energy
       R = 1.0Q0/(ALF+1)
       RR = -ALF*SP*AA/(2*(ALF+1)*(ALF+1)*PA)
       R = R+RR
       DO II = 2 , 20
         RR =  RR*((1-II*(ALF+1))/((II+1)*(ALF+1)))*(SP*AA/PA)
         R = R+RR
       END DO
       U = UA*(1+SP*AA/PA)**(1/(ALF+1))+SQ*AA*(1-R) ! inner energy
       T = U/(ALF*N*K)    ! temperature
       V = N*K*T/P     ! volume
     END IF
! volume and heat flow are given as functions of time in input
     IF((CHO.GT.1.5).AND.(CHO.LT.2.5)) THEN
       V = V+SV*AA   ! volume
       Q = Q + SQ*AA ! integral of added heat energy
       R = 1.0Q0/ALF
       RR = ((1-ALF)/(2*ALF*ALF))*SV*AA/VA
```

```
      R = R+RR
      DO II = 2 , 20
         RR =  RR*((1-II*ALF)/((II+1)*ALF))*(SV*AA/VA)
         R = R+RR
      END DO
         U = UA*(VA/(VA+SV*AA))**(1/ALF)+ALF*SQ*AA*(1+R/(1+SV*AA*R/VA))/
(ALF+1) ! inner energy
      T = U/(ALF*N*K)    ! temperature
      P = N*K*T/V     ! pressure
    END IF
! temperature and heat flow are given as functions of time in input
    IF((CHO.GT.2.5).AND.(CHO.LT.3.5)) THEN
      T = T+ST*AA  ! temperature
      Q = Q + SQ*AA ! integral of added heat energy
      R = 1.0Q0
      RR = SQ*AA/(N*K*TA)
      R = R+RR
      DO II = 2 , 20
         RR =  RR*((SQ*AA-(II-1)*N*K*ST*AA)/(II*N*K*TA))
         R = R+RR
      END DO
      U = ALF*N*K*T ! inner energy
      V = VA*((1+ST*AA/TA)**(-ALF))*R   ! volume
      P = N*K*T/V     ! pressure
    END IF
! pressure and volume are given as functions of time in input
    IF((CHO.GT.3.5).AND.(CHO.LT.4.5)) THEN
      P = P+SP*AA  ! pressure
      V = V + SV*AA ! volume
      T = P*V/(N*K)  ! temperature
      U = ALF*P*V ! inner energy
      X = UA-(PA+0.5Q0*SP*AA)*SV*AA
      Q = Q+U-X ! integral of added heat energy
    END IF
```

```
! pressure and temperature are given as functions of time in input
    IF((CHO.GT.4.5).AND.(CHO.LT.5.5)) THEN
      P = P+SP*AA  ! pressure
      T = T + ST*AA ! temperature
      V = N*K*T/P  ! volume
      U = ALF*N*K*T ! inner energy
      R = 1.0Q0
      RR = -0.5Q0*SP*AA/PA
      R = R+RR
      DO II = 2 , 20
        RR =  -RR*(II/(II+1))*SP*AA/PA
        R = R+RR
      END DO
      X = UA+N*K*TA*QLOG(1+SP*AA/PA)-N*K*ST*AA*R
      Q = QA+U-X  ! integral of added heat energy
    END IF
! volume and temperature are given as functions of time in input
    IF((CHO.GT.5.5).AND.(CHO.LT.6.5)) THEN
      V = V+SV*AA ! volume
      T = T + ST*AA ! temperature
      P = N*K*T/V  ! pressure
      U = ALF*N*K*T ! inner energy
      R = 1.0Q0
      RR = -0.5Q0*SV*AA/VA
      R = R+RR
      DO II = 2 , 20
        RR =  -RR*(II/(II+1))*SV*AA/VA
        R = R+RR
      END DO
      X = UA-N*K*(TA*QLOG(1+SV*AA/VA)+ST*AA)+N*K*ST*AA*R
      Q = QA+U-X  ! integral of added heat energy
    END IF
    W = W+(V-VA)*(P+PA)*5.0Q-1  ! integral mechanical work
    S = ALF*N*K*QLOG((ALF*P*V**((ALF+1)/ALF))/1Q0) ! gas entropy
```

```
      IF(ABS(Q).GT.ABS(U)/1.0Q11) THEN
         ETA = W/Q ! calculate efficiecy (gained mechanical work/added heat energy
       ETA = W/QMAX ! calculate efficiecy (gained mechanical work/added heat energy
         IF(ABS(ETA).GT.1.0) THEN
            ETA = ETA/ABS(ETA)
         END IF
      END IF
     if (I.EQ.1) THEN
        ETA1 = ETA
     END IF
     CALL MINMAX ! get to memory minimum and maximum values for the quantities
(for curve drawing purpose)
   END DO
!       WRITE(2,*) 'AIKA, V, VAPU, P, T, U,  W,   E  S  ETA'
      WRITE(2,1)  AIKA, V, VAPU, P/1Q5, TA-KAPS, U, W, Q, S, ETA
      WRITE(2,*)
      WRITE(3,2)  AIKA, V, P, T, U, W, Q, S, ETA, U0-(U-Q+W)
      LASK = LASK+1
   END DO
   CLOSE(UNIT = 3)
   OPEN(UNIT=3,FILE='IgPrintAPU.txt')
    DO I = 1, LASK
      READ(3,2)  AIKA, V, P, T, U, W, Q, S, ETA, ERRO
     if (I.EQ.1) THEN
        ETA = ETA1
     END IF
       WRITE(4,2)   AIKA, V/VM, P/PM, T/TM, U/UM, W/WM, Q/QM, S/SM, ETA/
ETAM, ERRO/U
    END DO
   WRITE(4,2)  AIKA, VMAX, PMAX/1Q5, TMAX, UMAX, WMAX, QMAX, SMAX,
ETAMAX
     WRITE(4,2)   AIKA, VMIN, PMIN/1Q5, TMIN, UMIN, WMIN, QMIN, SMIN,
ETAMIN
    print *, 'Hello World'
```

```
1 FORMAT(10(1X,G14.7))
2 FORMAT(10(1X,G14.7))


   CONTAINS


     SUBROUTINE LAUSEET(N) !reads dummy comment rows in the input file
     CHARACTER VIRKE
     INTEGER M,N
     DO 100 M=1,N
     READ(1,20) VIRKE
100    CONTINUE
20     FORMAT(A1)
     END SUBROUTINE LAUSEET

    SUBROUTINE MINMAX ! gets minimum and maximum values for the arguments
(used in drawing curves)
    IF (VMAX.LT.V) THEN
       VMAX = V
    END IF
    IF (PMAX.LT.P) THEN
     PMAX = P
   END IF
    IF (TMAX.LT.T) THEN
       TMAX = T
   END IF
   IF (UMAX.LT.U) THEN
     UMAX = U
   END IF
   IF (WMAX.LT.W) THEN
     WMAX = W
   END IF
```

```
IF (QMAX.LT.Q) THEN
   QMAX = Q
END IF
IF (SMAX.LT.S) THEN
   SMAX = S
END IF
IF (ETAMAX.LT.ETA) THEN
   ETAMAX = ETA
END IF
 IF (VMIN.GT.V) THEN
    VMIN = V
 END IF
 IF (PMIN.GT.P) THEN
  PMIN = P
END IF
 IF (TMIN.GT.T) THEN
    TMIN = T
END IF
IF (UMIN.GT.U) THEN
  UMIN = U
END IF
IF (WMIN.GT.W) THEN
  WMIN = W
END IF
IF (QMIN.GT.Q) THEN
   QMIN = Q
END IF
IF (SMIN.GT.S) THEN
   SMIN = S
END IF
IF (ETAMIN.GT.ETA) THEN
   ETAMIN = ETA
END IF
VM = MAX(ABS(VMAX),ABS(VMIN))
```

```
    PM = MAX(ABS(PMAX),ABS(PMIN))
    TM = MAX(ABS(TMAX),ABS(TMIN))
    UM = MAX(ABS(UMAX),ABS(UMIN))
    WM = MAX(ABS(WMAX),ABS(WMIN))
    QM = MAX(ABS(QMAX),ABS(QMIN))
    SM = MAX(ABS(SMAX),ABS(SMIN))
    ETAM = MAX(ABS(ETAMAX),ABS(ETAMIN))
  END SUBROUTINE MINMAX

    SUBROUTINE ARVI(U,A,LD,L,B)

!* linear interpolation with respect the first quantity of matrix B
!* U  = value
!* A  = argument
!* LD = dimension reservation
!* L  = real dimension
!* B(LD,2)  = matrix having values as a function of the argument

    REAL(16) B(LD,2),A,U
    REAL(16) AA, BB
    INTEGER LD,M,L,N

    DO M=2,L
      N=M
      AA = A
      BB = B(M,1)
      IF(AA.LT.BB) GOTO 100
    END DO
100   CONTINUE
    U=B(N-1,2)+(A-B(N-1,1))*(B(N,2)-B(N-1,2))/(B(N,1)-B(N-1,1))

    END SUBROUTINE ARVI

  end program IdealGas
```